Animal Research and Ethical Conflict

Mary T. Phillips   Jeri A. Sechzer

Animal Research and Ethical Conflict

An Analysis of the Scientific Literature: 1966–1986

Springer-Verlag
New York Berlin Heidelberg
London Paris Tokyo

Mary T. Phillips
34 Morton Street, 2-C
New York, NY 10014
USA

Jeri A. Sechzer
Department of Psychiatry
Edward W. Bourne Behavioral
Research Laboratory
New York Hospital–Cornell
University Medical Center
White Plains, NY 10605
USA

Library of Congress Cataloging-in-Publication Data
Phillips, Mary T., and Sechzer, Jeri A.
Animal research and ethical conflict.
Bibliography: p.
1. Animals, Treatment of. 2. Animal experimentation—
Moral and ethical aspects. I. Phillips, Mary T., 1945–
II. Sechzer, Jeri A., 1930– . III. Title.
HV4708.P55 1989 179'.3 88-36241
ISBN-13:978-1-4612-8181-8

Printed on acid-free paper.

Media conversion by Impressions, Inc., Madison, Wisconsin.

9 8 7 6 5 4 3 2 1

ISBN-13:978-1-4612-8181-8 e-ISBN-13:978-1-4612-3620-7
DOI: 10.1007/978-1-4612-3620-7

Acknowledgments

We are grateful to all those who reviewed early drafts of this volume for contributing their expertise in improving this work. In particular, we are indebted to Bruce H. Ewald, MA, DVM (CIBA-GEIGY Corporation); Phyllis B. Grodsky, PhD (Memorial Sloan-Kettering Cancer Center); D. Richard Knauff, DVM (Wyeth Laboratories); and Harry C. Rowsell, OC, DVM, PhD (Canadian Council on Animal Care). All four reviewers contributed numerous ideas, corrections, and suggestions, but we thank Dr. Ewald especially for the suggestion that we include the complete text of the Animal Welfare Act as amended in 1985; Dr. Grodsky for her review of the sociocultural background; Dr. Knauff for his critique of the chapter on the current American legislation; and Dr. Rowsell for his comments on the international section.

We also wish to thank William Thackeray for preparing the graphs and Peter G. Menegas for photographing them for reproduction.

Our gratitude is expressed to the following organizations for financial support provided to J.A.S. for this work: Bristol-Myers Company; Carter-Wallace Company; Colgate-Palmolive Company; Dow Chemical Company; Exxon Education Foundation; Merck, Sharpe and Dohme Research Laboratories; Procter and Gamble Company; Richardson-Vicks, Incorporated; Schering Foundation; and the Texaco Foundation.

Contents

Introduction

This volume grew out of a project to conduct a survey of the scientific literature of the United States for items addressing the ethical and humane issues of animal experimentation. The original purpose of the survey was to provide background information for the development of a set of guidelines for the use of animals in research then in preparation by The New York Academy of Sciences' Ad Hoc Committee on Animal Research.[1] We soon realized, however, that the survey constituted a valuable resource in its own right. In this book we present the results of the literature survey along with relevant information about its legislative and historical context. We hope that this material will be helpful to biomedical researchers grappling with animal welfare problems as well as to social scientists interested in tracing the evolution of the 20th-century phase of the animal research controversy.

The time span covered by the survey was selected to match the 20-year period between the passage of the first federal Animal Welfare Act in 1966 and the most recent legislation regulating animal research, which took effect during 1986. Our analysis examines this transitional period through the pages of the scientific literature. During these 2 decades, animal research in the United States was transformed. Activist animal protection organizations catapulted troublesome ethical questions to national prominence, throwing the scientific community[2] into turmoil. Our analysis reveals how scientists in the United States responded to criticisms from the public and how they treated animal welfare issues during this troubled time.

The results of the literature survey are presented in three ways: (1) in an annotated bibliography containing over 300 entries, arranged chronologically by year and alphabetically by author within each year; (2) in a quantitative analysis of items found, showing how scientific interest in the subject has waxed and waned since 1966; and (3) in a qualitative analysis of themes and viewpoints expressed.

In order to avoid creating the impression that the scientific literature exists in a social vacuum, we have included chapters outlining the historical background of the controversy, legislative developments in the United States, and the regulatory approaches taken by other countries. However, we have not attempted to provide a comprehensive discussion of these aspects of the subject.

The historical dimension of the debate, discussed briefly in Chapter 1, has received serious scholarly treatment elsewhere. Richard D. French's authoritative study of science and antivivisectionism in Victorian England, published in 1975, has already become a classic.[3] Other book-length works include those by James Turner[4] and Coral Lansbury,[5] and a volume edited by Nicolaas A. Rupke.[6] We refer readers to these authors for more thorough discussions of the history of the conflict between those who would protect animals and those who would defend science.

A more ambitious work would also consider the philosophical arguments advanced by Peter Singer, Tom Regan, and many others who have thought and written about the place of animals in our human moral world.[7] We have not presented the point of view of nonscientists, not because their views are unimportant or uninteresting, but because such an enterprise lies beyond the scope of this modest effort. We have also omitted that segment of the scientific community whose opinions might not be represented in the published literature. In short, we did not attempt to write the definitive study of the contemporary controversy over the use of animals in scientific research—but we hope to have laid the groundwork for such a project.

However, this study does represent the first attempt, to our knowledge, to gauge scientific attitudes about animal research objectively, using systematic and reproducible methods. In spite of certain *caveats* that must be kept in mind in the interpretation of these data, the results present a far more accurate picture of the attitudes of the scientific community than do the usual impressionistic and anecdotal assessments.

Another contribution of this book is the inclusion of appendices containing the full texts of the latest landmark legislation affecting animal research. Neither the Animal Welfare Act as amended in 1985 (Appendix A) nor the Health Research Extension Act of 1985 (Appendix B) is available elsewhere in easily accessible form.[8] We also are pleased to be able to include the Interdisciplinary Principles and Guidelines for the Use of Animals in Research, Testing, and Education prepared by the New York Academy of Sciences' Ad Hoc Committee on Animal Research (Appendix C). The guidelines, the development of which had provided the original impetus for our literature survey, were completed almost simultaneously with the manuscript for this book. The legislation and guidelines are relevant to the text, but apart from that, we expect that they will be of considerable practical value to scientists, to members of institutional animal care and use committees, and to students of the animal welfare movement.

The question of how—and if—we can ethically use animals for human ends tends to arouse intense emotions among participants in the debate. Yet readers will be disappointed if they hope to find in this book a strong partisan position that aligns us firmly with either side. We set out neither to defend nor to condemn, but to understand.

Notes

(The notation *[AB]* after a reference indicates that it is included in the Annotated Bibliography following Chapter 7.)

1 This committee was chaired by Jeri A. Sechzer during the development of the project and the preparation of the guidelines.

2 Our use of the term *scientific community* throughout this book represents a shorthand way to refer to the scientists whose views and opinions are under scrutiny here. By using this convenient term, we do not mean to endorse the theoretical view developed by Warren Hagstrom and others, which posits a group unified by shared norms. (See Hagstrom W [1965] *The scientific community* New York: Basic Books.) The question of whether scientific researchers form a community in this sense is beyond the scope of this essay, although the deep divisions on some issues uncovered in this survey might support the opposing viewpoint. For a brief discussion of this subject by a scholar who rejects the emphasis on consensus implicit in the concept of scientific community, arguing instead that controversy is an integral part of the scientific process, see Everett Mendelsohn (1987) The political anatomy of controversy in the sciences (In: Engelhardt HT Jr, Caplan AL [eds], *Scientific controversies.* Cambridge: Cambridge U Press, pp 93–124). Mendelsohn's paper is particularly pertinent to the present discussion because of its examination of the controversy over animal experimentation in Great Britain in the 19th century.

3 French RD (1975) *Antivivisection and medical science in Victorian society.* Princeton, NJ: Princeton U Press.

4 Turner J (1980) *Animals, pain and humanity in the Victorian mind: Reckoning with the beast.* Baltimore, MD: Johns Hopkins U Press. Turner takes a more psychological approach to the subject than does French, and he includes more information about developments in the United States.

5 Lansbury C (1985) *The old brown dog: Women, workers, and vivisection in Edwardian England.* Madison, WI: The U of Wisconsin Press. The author's focal point is a series of riots that took place in London in 1907 over a statue that had been erected by antivivisectionists in memory of a dog subjected to medical experiments at the University of London. Lansbury argues that working-class men (on this occasion, at least) and middle-class women were passionate supporters of antivivisection because they saw in vivisected animals symbols of their own oppression. The book contains a highly original discussion of parallels between vivisection, turn-of-the-century gynecological practices, and pornographic literature of the period.

6 Rupke NA (ed) (1987) *Vivisection in historical perspective.* London: Croom Helm. The 13 essays in this book focus on the historical origins of the vivisection controversy in western Europe and, in a paper by Susan E Lederer, the United States (The controversy over animal experimentation in America, 1880–1914. In: Rupke, op cit, pp 236–258).

[7] See, for instance, Singer's enormously influential *Animal liberation: A new ethics for our treatment of animals* (New York: New York Review/Random House, 1975; Avon, 1977). Other books by Singer bearing directly or indirectly on this subject include *The expanding circle: Ethics and sociobiology*. New York: Farrar, Straus & Giroux, 1981; and (ed) *In defence of animals*. New York: Blackwell, 1985. See also Singer's contribution to the entry on animal experimentation in the *Encyclopedia of bioethics* (Philosophical perspectives [1978] In: Reich WT [ed] *Encyclopedia of bioethics*, vol 1. New York: Free Press, pp 79–83) [AB]; and the following articles: Killing humans and killing animals (1979) *Inquiry 22*,1–2:145–156); Animal liberation (1980) In: Rundle JU (ed) *Science, technology and ethics*, Occasional Paper No 3 in the Connecticut Scholar series. Middletown, CT: Connecticut Humanities Council, pp 70–77; and The Oxford vegetarians—A personal account (1982) *Int J Stud Anim Problems 3*,1:6–9.

Books by Tom Regan include *The struggle for animal rights* (Clarks Summit, PA: International Society for Animal Rights, 1987); (ed) *Animal Sacrifices: Religious perspectives on the use of animals in science* (Philadelphia: Temple U Press, 1986); *The case for animal rights* (Berkeley, CA: U of California Press, 1983); *All that dwell therein: Essays on animal rights and environmental ethics* (Berkeley, CA: University of California Press, 1982); and a volume edited with Peter Singer, *Animal rights and human obligations* (Englewood Cliffs, NJ: Prentice-Hall, 1976). For articles by Regan, also see The moral basis of vegetarianism (1975) *Canadian J Phil 5*,2:193; Feinberg on what sorts of beings can have rights (1976) *Southern J Phil* (winter):485–498; McCloskey on why animals cannot have rights (1976) *Phil Quarterly 26*:251–257; Frey on interests and animal rights (1977) *Phil Quarterly 27*: 335–337; Narveson on egoism and the rights of animals (1977) *Canadian J Phil 7*:179–186; Animal rights, human wrongs (1989) *Environmental Ethics 2*,2:99–120; Animal rights and animal experimentation (1981) In: Basson MD (ed) *Rights and responsibilities in modern medicine: Progress in clinical and biological research*, vol 50. New York: Alan R. Liss, pp 69–83 [AB].

[8] The Health Research Extension Act is available in a pamphlet that also includes the PHS Policy on Humane Care and Use of Laboratory Animals (Bethesda, MD: Office for Protection from Research Risks, NIH, 1986). To our knowledge, the Animal Welfare Act is not available, even in pamphlet form. Official versions of the 1985 amendments, without the rest of the text, are published as Subtitle F of Title XVII of the Food Security Act in *United States Statutes at Large* (PL 99–198, 99 Stat 1645–1650, Washington, DC: U.S. Government Printing Office). The *United States Code* (Washington, DC: U.S. Government Printing Office) and the *United States Code Annotated* (St Paul, MN: West Publishing Co) contain the complete text, but as of this writing, these publications have been updated only through the 1976 amendments. The 1986 supplement to the 1982 edition of *USCA* contains the 1985 amendments, but refers readers to the main (1982) edition for the text of sections not changed by the 1985 amendments.

CHAPTER 1

Animal Research and Antivivisectionism: Historical Antecedents[1]

Early Events in Europe

The use of living animals for experimentation is documented as early as the second century AD. The Greek physician Galen (Claudius Galenus, AD 129–c 200) is considered the first to use the method of vivisection for scientific investigation. Galen was born in Greece and at about the age of 30 moved to Rome, where he gained great fame as a physician. During this period, autopsies were not permitted in Rome, and experiments on living animals became necessary.[2] Galen performed experiments on pigs and monkeys, injuring, destroying, or removing various organs to judge their function. He also used animals to study the effects of spinal cord sections, chest perforations, and arterial ligation.

The British pharmacologist Sir W.D.M. Paton, who has written extensively on legal and humane issues of animal experimentation, considers Galen as possibly the first significant animal experimenter.[3] Paton explores this by describing one of Galen's experiments, which showed that arteries contained blood, not air. Paton remarks that, although this fact should have been obvious, in the second century AD it would have been rare to see an artery in life without deliberate vivisection. After death, arteries contract and empty, while veins remain full. (Although the pulse was recognized, the understanding of its meaning and direction was delayed for several centuries.) The simple experiment that Galen carried out was to tie an artery in two places and then to make an opening between the two ligatures to reveal its contents. Thus, only with an experiment in life and the method of intervention could this simple discovery be made.

Using the method of Galen, eminent vivisectors such as Graaf, Harvey, Aselli, Pecquet, and Haller would make the practice of vivisection of animals a habitual, indispensable technique for the study of physiology and medicine.

Until the 18th century, there was no apparent moral concern expressed publicly about vivisection. Dissecting a living animal posed no ethical problem for Descartes since for him animals were machines, incapable of language. Descartes, as well as other thinkers of the time, considered the possession of language as the only evidence that one can think, feel, and reason, and he denied this capability in animals.[4] Francis Bacon argued that the dissection of living animals "notwithstanding the dissimilitude of their parts to human, may, with the help of a little judgement," advance our knowledge of human functioning.[5] Yet, in the next century, discussion would begin in England about the justification for animal experimentation. Samuel Johnson's writings in 1758 accurately anticipated much of the antivivisection movement in that country.[6]

By the 19th century, vivisection in England was under attack. Antivivisection and animal protection groups were organized by physicians, scientists, and the general public. In 1824 the "Society instituted for the purpose of preventing cruelty to Animals"[7] was established (and soon became the Royal Society for the Prevention of Cruelty to Animals [RSPCA]), with the stated purposes of publicizing the principle of kindness to animals, enforcing existing anticruelty laws, and passing new ones. As its membership increased, the RSPCA gained much influence in Britain. In 1857 the society turned its attention to issues of animal experimentation, joining with other groups to engage in purposeful activities to abolish "painful" animal research.[8]

By 1870 the antivivisection movement in England had become a raging controversy, with bills read in Parliament amid numerous demonstrations against animal research. That same year, the British Association for the Advancement of Science formed a committee to develop guidelines for conducting physiological experiments, to consider steps to minimize suffering, and to discourage experiments on animals that were not clearly legitimate. The report submitted by the committee in 1871 contained four parts:

i. No experiment that can be performed under the influence of an anaesthetic ought to be done without it.[9]

ii. No painful experiment is justifiable for the mere purpose of illustrating a law or fact already demonstrated; in other words, experimentation without the employment of anaesthetics is not a fitting exhibition for teaching purposes.

iii. Whenever, for the investigation of new truth, it is necessary to make a painful experiment, every effort should be made to ensure success, in order that suffering inflicted may not be wasted. For this reason, no painful experiment ought to be performed by an unskilled person with insufficient instruments and assistance, or in places not suitable to the purpose, that is to say, anywhere except in physiological and pathological laboratories, under proper regulations.

iv. In the scientific preparation for veterinary practice, operations ought not to be performed upon living animals merely for the purpose of obtaining greater operative dexterity.

Signed by: M.A. Lawson, Oxford
G.M. Humphry, Cambridge
John H. Balfour, Edinburgh

Arthur Gamgee, Edinburgh
William Flower, Royal College of Surgeons, London
J. Burdon Sanderson, London
George Rolleston, Secretary, Oxford[10]

These recommendations presaged the Cruelty to Animals Act of 1876 as well as concerns that are in considerable evidence today, not only in Britain, but in other European countries and in the United States.

Queen Victoria, strongly against vivisection, urged her prime minister, Disraeli, and her home secretary, Richard Cross, to bring some action to resolve the issue. In 1875 a royal commission was established to investigate vivisection in Britain, the amount of cruelty that might be taking place, and the best means of preventing it. The royal commission's report was the basis of a bill introduced in Parliament by Lord Carnarvon, which, after much lobbying and modification, became The Cruelty to Animals Act of 1876. The main points of the act were the following:

1. Any person wishing to perform experiments on living vertebrate animals must be licensed, and this license must be renewed yearly.
2. Experiments must be toward the end of alleviating suffering; any other purpose (e.g., to illustrate lectures) required certification.
3. Experiments on dogs, cats, horses, mules, and asses also required special certification.
4. Curare was not considered an anesthetic (and hence its use was banned).[11]
5. No one could be licensed as indicated except by permission of the home secretary.[12]

The act was implemented in 1876 and, with continuing additions and modifications to keep pace with new scientific developments, especially in pharmacology and behavioral science, remained in force until its revision in 1986. (See Chapter 7 for a summary of the 1986 British law.)

The Cruelty to Animals Act set into motion events that continued to unfold in England and foreshadowed developments in the United States during the last half of the 19th century. These roots would take hold and grow into the rhetoric of today's agitation against the use of animals for experimentation.

By the time the act was passed, it was watered down and did not really satisfy either side, but it was still restrictive. It required that scientists be licensed and that licensure be controlled by the home secretary. As a consequence, few licenses were issued, and the number of experiments and experimenters in Britain declined drastically. However, this did not end the controversy; there were more demonstrations against animal experimentation, new antivivisection groups were organized, new bills were introduced in Parliament, and a general campaign began to end all animal experimentation.

In 1882 the medical and scientific communities, concerned by the drastic decline in research and the hindrance to scientific progress, founded the Association for the Advancement of Medicine by Research (AAMR). It included an imposing membership: Sir William Jenner (president of the Royal College of Physicians), Sir James Paget, Sir William Gull, Dr. Fahrquharson, MP, Samuel Wilks, Joseph Lister, and Burdon Sanderson. The AAMR backed research of such scientists as Sir Charles Sherrington and J. Graham Brown.

The objective of the AAMR was to promote research in order to advance medical theory and practice in a variety of disciplines. This involved a careful scrutiny of the operation of the 1876 act, not with the idea of repeal, but to ensure its just implementation. The association began to play an important role in representing experimental medicine to the home secretary. By 1884, applications for licenses to experiment on living animals required the recommendation of the AAMR.

The administrative transfer of license decisions to the AAMR drastically altered the situation. The AAMR proceeded to issue licenses to large numbers of qualified investigators. The number of animal experiments in Britain increased significantly, and the experimental method soon became dominant. The antivivisectionists had not succeeded in abolishing or even critically limiting animal research after all. Instead, their intense dissatisfaction with the 1876 act had the effect of forcing the medical and scientific community in Britain to unite and protect science.[13]

Although much criticized by the antivivisectionist movement, this transfer of decision making on license applications from the home secretary to the AAMR remained in effect until 1913.[14] In that year, a second Royal Commission on Vivisection reported general satisfaction with the administration of the 1876 act, but recommended that the AAMR be replaced by an advisory body to the home office. Members of the advisory body were to be selected by the home secretary upon recommendations by the Royal Society and the Royal College of Physicians and Surgeons.

Notwithstanding the antivivisectionists' dissatisfaction with it, the law passed in 1876 did give the animals considerable protection. The Cruelty to Animals Act, the first law of its kind to be implemented on a national basis, had raised the consciousness of scientists, the government, and the general public about the need for the humane treatment of research animals. However, the battle continued, with the formation of many new antivivisection and animal welfare groups. The establishment of the animal *rights* movement (as opposed to animal *welfare*) awaited the writings of Peter Singer in 1975.[15]

By the early part of the 20th century, antivivisection activities had noticeably decreased. It has been suggested that the proven success of Pasteur's rabies vaccine (first tested in a human being in 1885) and the discovery of diphtheria antitoxin in 1894 were instrumental in demonstrating the value of animal research to the general public, thereby diminishing the effectiveness of the antivivisectionists' arguments.[16] Public sympathy for the antivivisection movement did not disappear completely, however, but fluctuated. Yet, until this last decade, protests against animal experimentation in England did not again approach the intensity of that in the late 19th century.

Events of the 19th century that led to the Cruelty to Animals Act of 1876 indicate the passionate feelings of many in England against the infliction of pain on animals, particularly domestic animals. This predominant concern was one reason why experimental medicine in England lagged behind crucial scientific advances across the channel in France, as well as in Germany, during this time. Another reason for this is thought to be the centralized structure of the universities and their domination by Oxford and Cambridge, where science meant "natural theology." The developing medical profession reflected this attitude. To this group, science meant observation, from which function could be inferred. Thus, throughout much of the 19th century, British medicine used anatomical inference as the basis of physiology.[17]

The distinction between the tradition of deductive inference from anatomy, which characterized British science early in the 19th century, and reliance on the vivisectional method, preferred in France, is illustrated by a famous scientific controversy. In 1822, 25 years before the discovery of anesthesia, François Magendie, a French physiologist, published the results of his now classical animal experiments on "cutting the anterior and posterior roots of the nerves arising from the spinal cord." After cutting the posterior (dorsal) roots to one or more limbs he noted the loss of sensation in the corresponding body part with movement unaffected. Magendie then cut the anterior (ventral) roots and found motor paralysis with no loss of sensation. In succeeding studies, Magendie repeated and varied these experiments several times on various species of animals. He concluded that the posterior roots seem particularly destined for sensibility, while the anterior roots seem to be especially connected with movement.[18]

The discovery that the dorsal and ventral roots have different functions—that the dorsal roots convey sensory signals and the ventral roots convey motor signals—has been considered one of the most important findings in the history of biology. This was the first evidence of localization of function in the nervous system and was the basis for the study of the spinal reflex by Sherrington, and thus led to much of our understanding of the complex physiology of the nervous system. In his correspondence with Cranefield about the Magendie experiments, H.E. Ackernacht remarked, "it is, after Harvey, probably the most momentous *single* discovery in physiology, and it had a more immediate influence on practical medicine than Harvey's discovery."[19]

The publication of Magendie's results gave rise to a bitter controversy between Magendie and Sir Charles Bell, an English anatomist. Bell claimed that Magendie had done no more than confirm and extend experiments that he had already performed in England 11 years earlier. The basis of Bell's claim was a pamphlet entitled "The Idea of a New Anatomy of the Brain," which he had distributed in 1811 to his circle of friends in England. In it, Bell describes his observation that injuring the dorsal roots did not affect the animal, but injuring the ventral roots resulted in convulsing the corresponding part of the animal's body.[20] He concluded that the dorsal and ventral roots have different functions, but erroneously supposed that the ventral roots subserve motion and sensation via the cerebrum and that the dorsal roots "belong to the cerebellum."[21]

Because of the significance of Magendie's results, the dispute soon took on an international flavor that was to last over a century. How it came about involves details of how the experiments were conducted and the attitudes of each of the investigators toward vivisection.

A clue to Bell's failure to determine the function of the dorsal roots can be found in his concern about the ethical status of animal experimentation and his reliance on anatomical interpretation. Bell's method of discovery was deduction from anatomy, not experiment. He had performed only one experiment on a living rabbit, sectioning the dorsal roots. He stated at the time, "I was deterred from repeating the experiment by the protracted cruelty of the dissection. I therefore struck a rabbit behind the ear, so as to deprive it of sensibility by the concussion and then exposed the spinal marrow."[22] If the rabbit had not been stunned, Bell probably would have obtained positive evidence that the dorsal roots conveyed sensation. Bell and his associate Shaw had also made repeated statements that the precise functions of the spinal roots could not be shown in a living animal because the traumatic surgery, the breaking of the bones required to expose the spinal cord, injured the cord. Magendie apparently was aware of this problem and conducted his initial experiments using 6-week-old puppies in which the vertebral column was not yet ossified and thus could easily be separated to expose the spinal cord without injuring it.[23]

Bell had no hesitation in showing his hostility to animal experimentation. Experiments, which in those days preceded the discovery of anesthesia, were extremely distasteful to him. In his correspondence the following passage appears:

> I should be writing a third paper on the nerves, but I cannot proceed without making some experiments, which are so unpleasant to make that I defer them. You may think me silly, but I cannot perfectly convince myself that I am authorized in nature, or religion, to do these cruelties—for what?—for anything else than a little egotism or self-aggrandisement; and yet, what are my experiments in comparison with those which are daily done? and are done daily for nothing.[24]

According to French, the Bell–Magendie controversy clearly illustrates the "*essence* of the vivisectional method for the solutions of questions of physiology and the complications and difficulties inherent in its development. Bell's antipathy to vivisection and his attachment to deductive inference from anatomy characterized the early 19th century British tradition as surely as Magendie's relentless animal experiments and empiricism ... presaged the future of French physiology."[25]

Bell's campaign against Magendie belittled his contributions and, taking advantage of sentiment against animal experimentation, included attacking him for conducting experiments that were cruel, thus supplying additional fuel for the antivivisectionists.

There is little doubt that Magendie's numerous and radical experiments, conducted 25 years before ether was discovered, were extremely severe and caused the animals intense, agonizing pain and suffering. Magendie was not indifferent to this, but stated that he experimented on animals because he did not wish to experiment on humans. Claude Bernard, Magendie's assistant, tells of a Quaker who visited Magendie and said to him, "for it is said that thou performest experiments on living animals. I come to thee to demand of thee by what right thou actest thus and to tell thee that thou must desist from these experiments, because thou hast not the right to cause animals to die or make them suffer, and because thou settest in this way a bad example and accustomest thyself to cruelty."[26] Magendie told the Quaker that if his experiments did not serve humanity they would indeed be cruel, but to use animals in order to make discoveries useful to medicine did not merit such reproach.

This was not the only example that distinguished British science from that in France during the 19th century. Magendie, now considered the father of experimental physiology, was succeeded at the College of France by his assistant, Claude Bernard. Bernard was a brilliant scientist who, in the course of his life, illumined so many aspects of physiology that Pasteur described Bernard "not as a physiologist but as physiology itself."[27] Every phase of physiology to which Bernard addressed his inquiries was explored with innumerable animal experiments. But it is his classic work with curare that is most relevant to this chapter.

In 1841, during the second year of his internship at the College of France, Claude Bernard became Magendie's assistant. Although Bernard conducted many experiments in Magendie's laboratory, he sought to find a place for independent studies of the chemistry of nutrition. He was soon provided with working space by Theophile Jules Pelouze, an established chemist. In 1844 Pelouze gave Bernard several arrows tipped with the South American poison curare, used by Indians to hunt and fight. It was known that curare had an immediate effect: immobility followed by rapid death. The outstanding characteristic seemed to be its quietness, with the victims showing no evidence of suffering. Bernard wrote of such a description by the explorer, Waterton, "of the death of an Indian hunter whose poison tipped arrow, shot upwards into a tree, had missed a monkey in the branches and, falling, had pierced the hunter's arm. Resigned to his fate, the Indian removed the little bamboo box containing the poison which was suspended from his shoulder, and putting it on the ground with his bow and arrow, stretched himself beside them to await death, which came quickly." Waterton reported the same effects in animals: "Birds in flight, pierced by tiny, needle-sharp arrows launched from a blowpipe, seemed not to feel the wound, but after a moment dropped without a flutter."[28]

Bernard did not have a theory in mind when he began his work with curare in 1849. In his own words, "we experiment to bring to birth observations which may in turn bring to birth ideas."[29] Bernard was curious about the action of the poison, the quietness of the death, and the apparent lack of suffering by the victim. We now know that the lack of suffering and the peacefulness of death is simply an illusion. Curare suppresses or paralyzes motor activity by altering transmission of excitation from nerves to muscle fibers across the neuromuscular junction. The victim of curare then is deprived of neither sensation nor pain, but only the means of expressing these by speech or by movement. Death comes when the nerve-controlled respiratory muscles no longer function and the victim is asphyxiated.

Bernard, of course, did not know this when he began his studies of curare on unanesthetized animals. He carefully recorded the action of the poison on a laboratory animal, making observations analogous to those of Waterton:

> A puppy, pricked on the thigh with a poisoned instrument, scarcely noticed the wound; it ran and leapt as usual, but at the end of three or four minutes the animal lay down on its belly as if it were tired; it retained all its intelligence and did not seem to suffer at all; it was merely disinclined to move. Soon the dog put its head on the ground between its two front feet, as if it were still more tired and would like to go to sleep. Nevertheless, its eyes were still open and peaceful while its body collapsed; the animal was then completely paralyzed. Soon its eyes became dull, its respiratory movements ceased, and the animal was dead, 8 minutes after the poisoned prick.[30]

Bernard continued his studies on frogs, rabbits, and dogs. He soon concluded that curare did not act on the nervous system as a whole; that the sensory nerves were unaffected, while motor nerves were prevented from transmitting a normal stimulus to the muscles. He was fascinated by curare, looking at it as a tool for the pharmacologic dissection of nerve and muscle systems. To the very end of his life, Bernard continued to devise ways of using this poison experimentally.

Bernard's work with curare anticipated the notions of junctions and synapses and paved the way for theories of neuromuscular transmission and nerve conduction. The clinical use of curare and curare-like substances in anesthesia and in treatment of denervated skeletal muscle are all based on the principles that Bernard discovered with animal experiments.[31]

Just as the animals in Magendie's experiments suffered agonizing pain, the animals that Bernard subjected to curare also suffered. Early in his studies with curare, Bernard came to understand this and described what he felt was the *real* plight of the sufferer:

> Can we imagine a more horrible fate than that of an intelligence watching the progressive loss of all the organs ... which are destined to serve it.... Our imagination cannot conceive of anything more wretched than that sensitive beings, who feel pleasure and pain, should be unable to seek one and avoid the other. The suffering which the imaginations of poets have invented is produced in nature by the action of the American poison. We can even say that fiction falls short of reality.[32]

Why then did he continue with his experiments with curare? Bernard obviously did not lack sensitivity about them or about how nonscientists both in France and England viewed his work. He was, however, quite firm in the belief that his work was justified by the possible benefits to humanity. In his book, *An Introduction to the Study of Experimental Medicine,* a chapter on vivisection further explains Bernard's feelings. He asks the question,

> Have we the right to make experiments on animals and vivisect them?... I think we have this right, wholly and absolutely. It would be strange indeed if we recognized man's right to make use of animals in every walk of life, for domestic service, for food, and then forbade him to make use of them for his own instruction in one of the sciences most useful to humanity. No hesitation is possible; the science of life can be established only through experiment, and we can save living beings from death only after sacrificing others. Experiments must be made either on man or on animals. Now I think that physicians already make too many dangerous experiments on man, before carefully studying them on animals. I do not admit that it is moral to try more or less dangerous or active remedies on patients in hospitals, without first experimenting with them on dogs; for I shall prove, further on, that results obtained on animals may all be conclusive for man when we know how to experiment properly.[33]

Ether was discovered in 1847, and as soon as it was introduced in France, Bernard began to use it in the laboratory and encouraged others to do the same in order to prevent and minimize pain.[34]

Clearly, throughout most of the 19th century, French scientists were not as concerned about the infliction of pain on animals as were the British. During this period, when antivivisectionist activity was building in England, British scientists were forced to go to France if they wanted to experiment on animals. Joseph Lister, for instance, worked in France on animal experiments that led to the development of better sutures.[35] In France there was little antivivisectionist activity. British organizations tried to encourage antivivisectionism in France, but they were largely unsuccessful.[36] French scientists such as Flourens, Gley, and Paul Bert would follow Magendie and Bernard in the 19th century, using the method of vivisection to make further important discoveries in science and medicine.

Animal protection legislation did not occur in France until 1963. An important factor was that antivivisection was of little appeal to the Parisian literati, then the leaders of public opinion. Their indifference apparently inhibited the growth of the antivivisection movement in that country while it thrived successfully across the channel. In contrast to the French, the leaders of public opinion in England were the aristocracy and the upper classes, who had always had a soft spot for animals, particularly domestic ones. It was in these groups that antivivisection was born; it was championed by others who identified with them as the defenders of moral responsibility.

Antivivisectionists in Germany also failed to gain much of a following for their cause in that country, where (as in France) the "superstitious reverence for science" prevailed.[37] German social leaders were advocates of science, which was esteemed by all, but particularly by the poorly educated, who had unswerving and unlimited faith in the power of science. Sharing in the veneration of science were the members of the German press, who thereby made the task of the antivivisectionists difficult and unrewarding.

Thus, during this period, while British medical journals asserted physicians' freedom from vivisection, scientists in other countries were founding the principles of the new experimental medicine, which, as in the work of Magendie and Bernard, relied heavily on animal experiments. Yet, wherever antivivisection activity did occur, in Germany or elsewhere on the continent, it was modeled after the British example.[38]

Legislation in Europe Through World War II

Legislation regulating animal experiments was virtually nonexistent before 1930, except in the United Kingdom, including Ireland.[39] Just prior to or during World War II, legislation regulating the use of animals for experimentation was passed in Germany, Italy, Liechtenstein, and Sweden.

Germany (1930, 1933, 1934)

During the late 19th century, Germany was unified by Bismarck, but certain states such as Bavaria and East Prussia maintained their identity. According to a German scientist, Klaus Gartner, Bavaria had regulations concerning the use of animals as early as 1833.[40] However, official regulations concerning vivisection came about in Germany only as Adolph Hitler assumed power. In 1930 regulations were established in Prussia that permitted experiments on living animals only for purposes of important research, and then only under prescribed conditions with avoidance of superfluous experiments for demonstration purposes (film presentations were suggested as a substitute). It was reported that all serious-minded scientific investigators supported this ministerial order.[41]

On August 17, 1933, Hermann Göring, chairman of the Prussian cabinet, issued an order to be effective immediately that "vivisection of animals of whatsoever species [is prohibited] in all parts of Prussian territory." Persons who engaged in vivisection of animals would be punished by removal to a concentration camp.[42] In Bavaria, too, vivisection was prohibited. At the time these prohibitions were announced, it was generally understood that they were the prologue to a corresponding federal law for the protection of animals in all of Germany, the elaboration of which was already under way by the federal Ministry of the Interior and the federal Bureau of Health. However, it was carefully specified that necessary diagnostic and therapeutic tests using animals, such as insulin control and the diagnosis of renal tuberculosis, were permitted. The Minister of the Interior expressly stated that "consideration will be given to the requirements of science."[43]

On February 1, 1934, the government of Adolph Hitler enacted the new federal law pertaining to the protection of animals.[44] For the general protection of animals, the law stated that "unnecessary tormenting or rude treatment of animals is prohibited," and the specifics, including punishment for such treatment, were spelled out in detail. Relevant here are the regulations concerning experiments on living animals. In principle, the law prohibited scientists from performing interventions on living animals or applying treatments to them if such interventions were associated with noteworthy pain or damage. This prohibition covered all painful and operative experiments, including bloodless or other experiments. The term *experiment* comprised not only the first intervention or treatment, but also the entire course of an experiment extending over a long period of time.

These regulations were designed to eliminate unauthorized persons from participating in animal experiments. The Minister of the Interior had the authority to extend permission to certain institutions to conduct animal experiments if all the regulations were met. Any person found to undertake such experiments without permission was liable to a 6-month prison term in addition to a fine. The 1934 law specified that experiments were to be carried out only on the full responsibility of the scientific director of the institution or his specifically authorized representative, and only by properly trained persons, or under their direction, and then only with the avoidance of all unnecessary infliction of pain. The regulations also stated that only reliable persons who had completed a course in a university were eligible to become directors of research institutes. Experiments for research purposes were permitted to be undertaken only if they promised a definite result not yet confirmed by science, or would serve to clarify previously unresolved problems. Experiments were to be performed only under anesthesia (local or general) unless, in the opinion of the scientific director, the use of anesthetics was precluded by the purpose of the experiment, or "in case, by the application of anesthesia, an animal would be worried more" or its well-being would be more impaired by the anesthetic than by the intervention itself. Only in exceptional cases could a second experiment be performed on the same animal. In extensive interventions or in the event of severe pain, animals were to be killed immediately. Experiments on horses, dogs, cats, or monkeys could be carried out only when the objective of the research could not be obtained using other animals. Animal experiments for demonstration purposes were permitted only if pictures, models, prepared specimens, and films would not be sufficient. Reports had to be made of animal experiments, including a clear statement of their purpose.

There were exceptions to these provisions connected to the administration of justice, or for vaccinations and blood withdrawal for diagnostic purposes, or for the purpose of obtaining or testing serums and vaccines in accordance with recognized or approved procedures. Under these conditions too, if necessary, the animals were to be painlessly killed. Dr. Geise, then the administrator of the federal Ministry of the Interior, stated that, through these controlled regulations, it would be guaranteed that experiments on living animals would no longer be regarded by a large portion of the German population as cruel or immoral acts, but rather as an important link in the chain of research methods. The daily press was also "expected" to desist from publishing reports of experiments on living animals, and the medical press was "urged" to describe experiments so that it would be apparent that all possible consideration had been shown to the experimental animals.

The development and enactment of regulations to protect animals, especially in laboratory experiments, were, according to Seidelman, "a tragic irony of the satanic values of the Nazi regime.... Concomitant with the program to dehumanize Jews, the Hitler regime enacted legislation protecting other animals."[45] The 1934 law enacted by the Third Reich was so similar in principles and details to that of the British law of 1876 as to suggest strongly that Hitler had a selective admiration for British values—for the value of animal life, but not for human life. At the time that the Nazi law was implemented, it was, next to the British law, the most humane law in the world concerning the use of animals for experimentation. Hitler is said to have been an antivivisectionist who, out of concern for the safety of the animals at the Berlin zoo, ordered air raid shelters to be built for the animals so that they could be protected during the allied bombings.

After World War II, all laws enacted by the Nazi regime were abolished, and the Federal Republic of West Germany did not enact new laws concerning the use of animals in research until 1972. The postwar constitution affirmed the freedom of scientific inquiry without directly addressing the animal research issue.[46]

Italy (1931)

Mussolini, possibly under the influence of Hitler, passed an Animal Protection Law in 1931. It was amended in 1941. Under this law, experiments may be performed only by named, suitably qualified individuals in authorized institutes in which the director holds responsibility. The law provides for the right of inspection by medical and veterinary officers of provincial health authorities. Adequate anesthesia is required unless it will interfere with the purpose of the experiment. Only warm-blooded vertebrates are covered by the law, which has an annual reporting requirement. The law was and is poorly administered, and new laws have been under consideration.[47]

Liechtenstein (1936)

Liechtenstein passed an Animal Welfare Act in 1936, which prohibits vivisection. The law is still in place, but is currently under review.[48]

Sweden (1944)

Sweden's Protection of Animals Act (1944) established general conditions for animal welfare, transport, and public exhibition of animals, as well as regulations for the use of animals for scientific purposes. Animals were not to be used without permission of the Board of Agriculture for scientific research or education, diagnosis of disease, production of medicines, or any similar purposes when the use would involve animal suffering. Exemptions could be granted by the king. However, this restriction did not apply to universities, laboratories, or hospitals that were run by the state. Detailed regulations for the latter were contained in an edict of 1944, but were not implemented until January 1967. These regulations included the requirement that if the director of such an establishment were not a physician, veterinarian, dentist, or pharmacist, he or she had to be approved by the Board of Agriculture. This law was not amended until 1979.[49]

By the end of the 1960s, Denmark, Poland, France, and Luxembourg had added their names to the list of nations with provisions for the monitoring of animal experiments. Today, Austria, Belgium, Finland, Iceland, Japan, The Netherlands, Norway, Switzerland, and the United States have legislation for the regulation of animal experimentation. Regulation is also in effect in parts of Australia, and legislation is pending in Portugal. In Canada there is a voluntary peer review program as an alternative to legislation. There are several sets of international regulations, both legislative and voluntary. Many of these statutes are in part a legacy of the original British movement.[50] A country-by-country summary of post-World War II legislation around the world is given in Chapter 7.

Notes

(The notation *[AB]* after a reference indicates that it is included in the Annotated Bibliography following Chapter 7.)

[1] We thank The New York Academy of Sciences and Pergamon Press for permission to reprint parts of this and the following chapters (with minor alterations) from previously published works. The section on the Bell–Magendie controversy and on Claude Bernard is reprinted from Sechzer JA (1983) The ethical dilemma of some classical animal experiments. In: Sechzer JA (ed) *The role of animals in biomedical research.* Annals of The New York Academy of Sciences, vol 406. New York: The New York Academy of Sciences, pp 5–12. Parts of "Early events in Europe" (Chapter 1) and "Events in the United States from the 19th century through World War II" (Chapter 2) are reprinted from *Social Science and Medicine 15F*,1:13–17, Sechzer JA, Historical issues concerning animal experimentation in the United States, Copyright 1981, Pergamon Press plc.

[2] Torack RM (1982) Historical aspects of normal and abnormal brain fluids. *Arch Neurol 39*:197–201.

[3] Paton W (1984) The benefits of animal experiment. In: *Man and mouse: Animals in medical research.* Oxford: Oxford U Press, pp 42–43.

[4] Zola JC, Sechzer JA, Sieber JE, Griffin A (1984) Animal experimentation: Issues for the 1980s. *Sci Tech Human Values 9*,2:40 [AB].

[5] Ibid.

[6] French RD (1975) Animal experiment and humanitarian sentiment before 1870. In: *Antivivisection and medical science in Victorian society.* Princeton: Princeton U Press, pp 16–17.

[7] Ibid, pp 26–27.

[8] Historian Richard French describes the activities of this group and the social forces that were responsible for its development and increasing influence, and for stimulating the founding of similar organizations. (See French, ibid, pp 16–35).

[9] It is difficult to determine how widely anesthesia was used in animal experimentation at this time. For evidence on this point, science historian Stewart Richards examined the most widely known physiological textbook in use in 19th century England, John Burden Sanderson's *Handbook for the physiological laboratory* (1873). Richards found that about 15% of the experiments described in the handbook were both likely to entail pain and did not specify the use of any anesthetic, or specified the use of curare alone (Richards S [1987] Vicarious suffering, necessary pain: Physiological method in late nineteenth-century Britain. In: Rupke NA [ed] *Vivisection in historical perspective.* London: Croon Helm, pp 125–148; see also Richards S [1986] Drawing the life-blood of physiology: Vivisection and the physiologists' dilemma, 1870–1900. *Ann Sci 43*:27–56). Sanderson testified before the royal commission in 1875 that anesthesia was not mentioned in these experiments because its use was taken for granted, but another contributor to the handbook, Emanuel Klein, destroyed the cred-

ibility of that defense by testifying that he never used anesthesia except for convenience. Richards' examination of Sanderson's laboratory notebooks revealed that the latter sometimes used anesthesia and sometimes did not, with no explanation of the inconsistency.

10 French RD (1975) Experimental medicine in Britain. In: French, ibid, p 45; see also p 413.

11 This proviso stemmed directly from experiments with curare carried out in France by Claude Bernard.

12 An Act to Amend the Law Relating to Cruelty to Animals. (15th August, 1876) Public General Statutes 39 and 40, Victoria, Ch 77; Cruelty to Animals Act 1876, Guidance Notes. London: Research Defence Society.

13 French RD (1975) Administration of the Act and the A.A.M.R.. In: French, op cit, pp 210–219 [note 6]; Sechzer JA (1981) Historical issues concerning the use of animals in the United States. *Soc Sci Med 15F*:13–17 [AB].

14 French RD (1975) Op cit, pp 204–208 [note 6].

15 Singer P (1975) *Animal liberation.* New York: New York Review/Random House.

16 Ritvo H (1984) Plus ça change: Antivivisection then and now. *Sci Tech Human Values 19*,2:60–61 [AB].

17 French RD (1975) Experimental medicine in Britain. In: French, op cit, pp 36–37 [note 6].

18 Magendie F (1822) Expériences sur les fonctions des racines des nerfs rachidiens. *J Physiol Exper Path 2*:276–279.

19 Cranefield PF (1974) *The way in and the way out: François Magendie, Charles Bell and the roots of the spinal nerves.* Mount Kisco, NY: Futura Publishing Company, pp xiii–xiv.

20 Bell C (1966) *Idea of a new anatomy of the brain.* Winchester, for Dawsons of Pall Mall, London (a facsimile of the privately published edition of 1811): Warren and Son, Ltd, The Wykeham Press, pp 19–26.

21 Olmstead JMD (1944) *François Magendie.* New York: Schuman's, pp 103–110.

22 Ibid, p 116.

23 Cranefield PF (1974) Op cit, p 11 [note 19].

24 Gordon-Taylor G (1958) *Sir Charles Bell: His life and times.* Edinburgh and London: E and S Livingstone, Ltd, pp 109–112.

25 French RD (1975) Op cit, p 19 [note 6].

26 Dawson PM (1908) *A biography of François Magendie.* New York: Albert T Huntington, pp 53–55.

27 Olmstead JMD (1953) Claude Bernard (1813–1878). In: Haymaker W (ed) *The founders of neurology.* Springfield, IL: Charles C Thomas, pp 107–109.

28 Olmstead JMD, Olmstead EH (1952) *Claude Bernard and the experimental method in medicine.* New York: Henry Schuman, pp 75–77.

29 Ibid, pp 73–74.

30 Ibid, pp 76–77.

31 Fessard A (1967) Claude Bernard and the physiology of junctional transmission. In: *Claude Bernard and experimental medicine.* Cambridge, MA: Schenkman Publishing Company, Inc, pp 105–123.

32 Olmstead JMD, Olmstead EH (1952) Op cit, p 78 [note 28].

33 Bernard C (1957) *An introduction to the study of experimental medicine.* New York: Dover Publications, pp 101–103; translated into English by Henry Copley Greene (the original work was published in France in 1865). Bernard's concern

for the use of humans as subjects in research antedated, by more than 100 years, regulations for the protection of human subjects in research now in place in many countries.

[34] Although William Morton is credited with the discovery of ether in 1847, Crawford Long had provided evidence for the anesthetic qualities of ether on March 30, 1842, when he used it to excise an encysted tumor from the neck of a patient who had repeatedly put off the operation because of his fear of the pain involved. Long reported that when he informed the patient that the operation was over, the man was incredulous until Long presented the excised tumor. However, publication of Long's findings and his subsequent use of ether as an anesthetic in surgical procedures were delayed until 1849. See Comroe JH Jr (1978) Publish and/or perish. In: *Retrospectroscope: Insights into medical discovery*. Menlo Park, CA: Von Gehr Press, pp 75–77.

[35] Goodrich JE (1977) The first 100 years of antivivisection 1824 to 1924. *Mayo Clinic Proceedings 52*:257–259 [AB].

[36] There were some demonstrations against Magendie, and later Bernard, on the grounds of cruelty to animals. Bernard's wife and daughter, both antivivisectionists, carried out the most consistent attacks on Bernard, virtually banishing him from their household. See Zola JC et al, op cit, p 40 [note 4].

[37] French RD (1975) Anatomy of an agitation. In: French, op cit, p 235 [note 6].

[38] Also see French (op cit, pp 220–287 [note 6]) for a more detailed discussion of the social, cultural, and intellectual influences on attitudes toward animal experimentation in France and Germany.

[39] As part of the United Kingdom, Ireland has had legislation since the passage of the British Cruelty to Animals Act of 1876.

[40] Gartner K (1985) 100 years of animal welfare regulation and of experiments with animals in Germany, effects and reasons. Presented at the second CFN Symposium, The Ethics of Animal Experimentation. Stockholm, Sweden.

[41] Berlin: New regulations concerning vivisection. (1933) *J Am Med Assoc 101*,14:1087.

[42] Seidelman WE (1986) Animal experiments in Nazi Germany. *The Lancet 1*:1214.

[43] Berlin: New regulations concerning vivisection. (1933) Op cit [note 41].

[44] The law pertaining to protection of animals. (1934) *J Am Med Assoc 102*,7:551–552; Tierschutzgesetz, Vom 24, November 1933, enacted February 1, 1934.

[45] Seidelman WE (1986) Op cit [note 42].

[46] Personal communication to JAS from Dr. Jenny Remfry, Universities Federation for Animal Welfare (Potters Bar, UK) and Professor Stian Erichsen, National Institute of Public Health (Oslo).

[47] Hampson J (1987) Law relating to animal experimentation. In: Tuffery AA (ed) *Laboratory animals: An introduction for new experimenters*. London: Wiley, p 49; Taylor GB (1975) Animal welfare legislation in Europe. In: *Animals and the law*. Hertfordshire, England: Universities Federation for Animal Welfare, p 38.

[48] Hampson, ibid.

[49] Ibid, p 51; Taylor, op cit, p 39 [note 47].

[50] Rankin JD (1984) Monitoring of animal experimentation: Statutory controls. In: Bankowski Z, Howard-Jones N (eds) *Biomedical research involving animals*. (Proceedings of the XVIIth CIOMS Round Table Conference) Geneva: Council for International Organizations of Medical Sciences (CIOMS), p 140.

CHAPTER 2

The American Context: Background and Current Regulations

Events in the United States from the 19th Century Through World War II

In the United States, concerns about the use of animals for research have existed since the last part of the 19th century, but until 1985 the conduct of animal research here was virtually unaffected. Events that changed the course of research in this country can be traced in many ways to those in 19th century Britain, reviewed above. The strength of the animal protection movement today is not simply "a product of the expanded liberal sensibilities of the 1960s and 1970s [or] ... a sudden rise in consciousness and activism," but rather expresses the depth of its historical roots.[1] During the second half of the 19th century, American social activists began to agitate on behalf of animals in an echo of earlier developments in England.

The first *organized* humanitarian effort in the United States came about in 1866 when Henry Bergh founded the first American Society for the Prevention of Cruelty to Animals. These anticruelty societies were not concerned with the epizootics that occurred in cattle, horses, or hogs, or with the lack of shelters for livestock dying on the midwestern plains. Nor at this time were they concerned about the use of animals for research purposes, as in Britain. Instead, they wanted to do something about the cruelties that individuals practiced on animals in cities. By 1910 there were 131 societies organized solely for the protection of animals; today, there are thousands.[2]

In 1867 and again in 1880, antivivisection bills were presented to the New York State Legislature, but both failed to pass. Had either bill been enacted, all animal research in New York State would have been halted. Three years later, in 1883, the first Antivivisectionist Society was established in Philadelphia, with the object of preventing needless suffering of animals under the pretense of medical or scientific research.[3]

The antivivisection movement gained in strength and membership toward the end of the century. This change came about in part from the growing importance of experimental physiology, pathology, and other scientific disciplines that relied on the use of animals in the education of physicians. There were increasing efforts to have state legislatures and the Congress regulate, restrict, or abolish animal experimentation. Further attempts to pass antivivisection bills were made from 1896 to 1906 in Massachusetts and Pennsylvania, but these, too, failed.

Antivivisectionist issues were strongly contested during this period by physicians and scientists who lobbied successfully against restrictive legislation. Benison, in his article, "In Defense of Medical Research," has described these efforts in some detail. In 1896 Professor Henry Bowditch of Harvard obtained the support of the Massachusetts Medical Society in lobbying against an antivivisection bill. Four years later, Dr. William Welch of Johns Hopkins University was joined by Drs. William Osler, William Keen, Robert Hare, George Sternberg, and other prominent physicians and teachers to oppose a bill that would regulate animal experimentation in the District of Columbia. In 1908, following an antivivisectionist attack against Rockefeller University, the American Medical Association organized a Special Defense Committee in Support of Medical Research. Professor Walter B. Cannon served as the chairman of the committee for 18 years. Benison credits Cannon with "direct[ing] the struggle against antivivisection in the United States," while affirming the right of medical research and experimentation. Cannon developed a code of laboratory procedures for medical schools and research institutes, and he motivated leaders in various medical fields to write papers on the importance of animal research in medical and surgical practice.[4]

Current American Regulations

After World War II, animal use in science increased, and with this increase came the realization that steps needed to be taken to ensure proper animal acquisition. In 1962 the National Institutes of Health (NIH) contracted with the National Academy of Sciences' Institute of Laboratory Animal Resources (ILAR) to develop the *Guide for Laboratory Animal Facilities and Care*.[5] Soon thereafter, the 1966 Animal Welfare Act became the first federal statute to deal with the proper acquisition of animals and to make an attempt to address the humane treatment of animals intended for scientific research.[6] Twenty years after that tentative effort, major legislation was enacted that would have a significant effect on the conduct of animal experimentation in this country. New Public Health Service (PHS) regulations on humane care and use of laboratory animals were implemented during 1986, and in December of that same year, important amendments to the Animal Welfare Act were also implemented. This marked the end of a long and troubled transition for the United States, from the ambiguous and limited governmental policy initiated in 1966 to a highly focused set of regulations applicable to a major proportion of research institutions in this country.[7]

The formulation of the new policy emerged after a 20-year period of debate over animal use in the United States. These two decades, from 1966 to 1986, can be seen as an era of historical transition from which the United States has come forth to take a prominent role along with many other countries in evincing concern about the ethical use of animals for experimentation. The prolonged turmoil paralleled by changes in the way scientists view the issue is clearly reflected in the scientific literature of the period, as the following chapter will illustrate.

The significance of this era lies in the changing role of science and technology in the United States, and changing perceptions of, and attitudes toward, that role on the part of both the scientific community and the general public. Equally important is the increased involvement of the government in funding and conducting scientific research. This has been accompanied by a movement toward the articulation of a set of professional ethics and values in science.

Changes in animal research policy over the past 2 decades have taken place against a background of heightened public awareness of many moral and ethical problems: questions involving nuclear arms and energy, genetic engineering, abortion, artificial insemination, human rights, and human experimentation. Animal research is one such issue that has now assumed international proportions. Few topics have been able to generate as much emotional passion and fervor on a sustained basis as have concerns about animal use in biomedical and behavioral research. Nor have many issues engendered such mistrust between the scientific community and the public.

Tactics of the animal protection movement have often been directed toward the goal of governmental regulation of animal experimentation. Yet our legislative system does not make this easy. Laws concerning animal experimentation in the United States are limited by the fact that, under the Constitution, Congress is not given the specific power to regulate the use of animals in research. Such power must come indirectly through the commerce clause as a condition of receiving funds to carry out federally sponsored research, or laws may be passed by individual states.[8] The complexity of the process by which proposed laws are passed in the United States and the time this takes tend to preclude hastily conceived legislation, but at the same time can frustrate the desire to translate public sentiment speedily and directly into law. This type of hindrance can amplify the protests of social activists in their efforts to produce and hasten desired results.

In the course of the period under discussion, public attitudes have reflected a growing awareness of the social and ethical costs of scientific progress. The role of government in scientific research has grown, bringing a change in the politics of science and diminished prestige for scientists.

Mistrust of uses to which scientific knowledge is applied pervades public thought. The young have become increasingly suspicious of science and ambivalent about the scientific method. To make matters worse, as far as animal research is concerned, there has also developed a sympathy for the notion of animal "awareness"—the possibility that animals have consciousness and subjective feelings. In addition, challenges to the authority of established institutions, both political and scientific, have heightened the perception that science should be held accountable to the lay public.[9]

In assessing the social and political maelstrom focusing on the animal experimentation issue, we acknowledge the sincerity and compassion that underlie the activities of even the most radical of the animal protection groups. Some of these activities have startled and angered the research community, especially when they involved illegal break-ins, theft of experimental animals, and destruction of laboratory equipment. In several well publicized cases, however, animal protectionists succeeded in exposing serious violations of the Animal Welfare Act and PHS policy, resulting in suspension of federal funding for the research involved.[10] Hence, regardless of our view of the methods taken by the extremists, we must recognize the important issues that have been raised. In the words of the ethical philosopher Arthur Caplan, these issues revolve around the following questions: "1) is it morally legitimate to conduct research on animals?; 2) if it is morally permissible to utilize animals for research purposes, what moral responsibilities must be discharged by those engaged in research?"[11]

The Laboratory Animal Welfare Act enacted on August 24, 1966, did not attempt to answer these questions. (The title was changed in 1970 to The Animal Welfare Act.) Its primary objectives were to protect the owners of dogs and cats from theft of their pets, to prevent the sale or use of stolen dogs or cats for purposes of research or experimentation, and—last—"to insure that certain animals intended for use in research facilities are provided humane care and treatment."

Among the provisions of the act were requirements for licensing dealers and for registering research facilities with the Animal and Plant Health Inspection Service (APHIS) of the United States Department of Agriculture (USDA). Regulated research facilities were defined as "any school, institution, organization, or person that uses or intends to use dogs or cats in research, tests, or experiments, and that (1) purchases or transports dogs or cats in commerce, or (2) receives funds under a grant, award, loan, or contract from a department, agency, or instrumentality of the United States for the purpose of carrying out research, tests, or experiments." The Secretary of Agriculture was empowered to establish and promulgate standards for humane handling, care, treatment, and transportation of animals by dealers and research facilities. Such standards included minimum requirements with respect to housing, feeding, watering, sanitation, ventilation, shelter from extremes of weather and temperature, separation by species, and adequate veterinary care. The secretary was also directed to establish rules for the inspection of premises and for the required record keeping by licensed dealers and registered research facilities of the purchase, handling, or sale of cats and dogs.

A major limitation of the 1966 act lay in the fact that it covered only those facilities that used dogs and cats. The act regulated some additional species (nonhuman primates, guinea pigs, hamsters, and rabbits), but this was a moot point for facilities that were exempt because they used neither dogs nor cats. The exclusion of rats and mice—the most frequently used species in research facilities—from the act's protection further reduced its applicability.

The legislators who drafted the 1966 Animal Welfare Act apparently never intended to consider the purposes for which the animals were used in research or the treatment they received in experimental procedures. The legislation specifically stated that "Nothing in this Act shall be construed as authorizing the Secretary [of Agriculture] to promulgate rules, regulations, or orders for the handling, care, treatment, or inspection of animals during actual research or experimentation by a research facility as determined by such research facility." The regulation of animal use *during* experiments had not yet become a major issue in the United States, although it was beginning to be discussed in Europe. Some states began to change their laws in order to restrict the use of pound animals by research laboratories, but such laws and regulations still did not confront directly the question of the use and treatment of animals in research, testing, or education.[12]

When the Animal Welfare Act was introduced in Congress, it produced relatively little discussion in the scientific literature.[13] This was in marked contrast to the heated disputes between scientists and animal protectionists in hearings held to consider the proposed legislation on animal experimentation in the 1980s.[14] Although the message of the 1966 legislation was essentially cautionary, it gave credence to the idea that the issue was ready to undergo critical review in the United States by a variety of concerned groups.[15]

Since 1966 the Animal Welfare Act has been administered by the USDA, which registers research facilities and requires them to keep records of the regulated species they use. Enforcement of the act is carried out primarily through inspection by the USDA's Animal Plant and Health Inspection Service (APHIS), but the complaint is frequently heard that APHIS's staff and budget are inadequate for the task of conducting appropriate inspections and assuring compliance with the provisions of the act. Increasing funding of the USDA to facilitate inspections has recently become an important concern for both scientists and animal welfare groups. Nevertheless, federal funding for enforcement of the Animal Welfare Act dropped over the past few years, from $4.9 million for 1982 to $4.628 million for 1986.[16] This compares poorly with the figure of $10 million needed for the task, according to a report issued by the National Research Council.[17]

The first amendments to strengthen the Animal Welfare Act were passed in 1970 (PL 91-579). These amendments brought more research facilities under the act's coverage by including all facilities that use "live animals," rather than restricting the definition to those using cats or dogs. Whereas the 1966 act had defined *animal* to mean dogs, cats, nonhuman primates, guinea pigs, hamsters, and rabbits, the 1970 amendments added to this definition the phrase, "or such warm-blooded animal, as the Secretary may determine is being used, or is intended for use, for research, testing, experimentation, or exhibition purposes, or as a pet." Specifically excluded were farm animals, such as "horses not used for research purposes" and poultry and livestock used for food or fiber. Since mice and rats were not mentioned, it would appear that the definition was now broad enough to include these ubiquitous laboratory creatures. According to a report written by the Office of Technology Assessment, this was indeed the intent of Congress.[18] In practice, however, mice and rats have never been considered regulated species, and in 1977 the Secretary of Agriculture promulgated regulations that specifically excluded rats, mice, birds, horses, and other farm animals from the act's coverage.[19] Because of the exemption of mice and rats, which is still in effect, it has been estimated that only 4% to 5% of animals used in research are covered by the act.[20]

The 1970 amendments also elaborated on the concept of *adequate veterinary care*, a phrase that had been merely listed as a requirement under the standards of the 1966 act. Now it was stipulated that adequate veterinary care include "the appropriate use of anesthetic or tranquilizing drugs, when such use would be proper in the opinion of the attending veterinarian of such research facilities." The early, yet weak, attempt to address the issue of animal pain passed virtually unnoticed at the time, due perhaps to the proviso that "nothing in this Act shall be construed as authorizing the Secretary to promulgate rules, regulations, or orders with regard to design, outlines, guidelines, or performance of actual research or experimentation by a research facility...."

This "hands-off" stance toward research procedures is still written into the act to reassure scientists that their professional autonomy cannot be usurped by the federal government. However, the effect was attenuated somewhat by another new provision contained in the 1970 law: a requirement that research facilities covered by the act submit an annual report to the USDA. The report must "show that professionally acceptable standards governing the care, treatment, and use of animals, including appropriate use of analgesic and tranquilizing drugs, during experimentation are being followed by the research facility during actual research or experimentation." Regulations promulgated by the Secretary of Agriculture added the requirement that the report must include the approximate number of each species on which such experiments, tests, or research was conducted, and it must specify how many animals of each species received pain-relieving drugs. An explanation must be included for any painful experiments in which the animals did not receive such drugs.[21] These annual reports are public record, available by request under the Freedom of Information Act.

At the same time, the 1970 amendments provided that, not later than March of each year, the Secretary [of Agriculture] submit to the President of the Senate and the Speaker of the House of Representatives a comprehensive and detailed written report with respect to "1) the identification of all research facilities, exhibitors, and other persons and establishments licensed...; 2) the nature and place of all investigations and inspections made...; and 3) recommendations for legislation to improve the administration of the Act or any provisions thereof...." This report, published annually by the USDA, also contains summaries of the information provided in the facilities' individual reports. The document is thus a source of official, but incomplete, data on animal use in the United States. Only regulated species are required to be reported, and unregistered facilities are not required to file reports at all. Consequently, it is still difficult to determine how many animals are used in the United States annually. The total number of *regulated* animals used in 1986, according to the USDA report to Congress for that year, was 1,778,403.[22] Estimates of the actual number of animals used (including mice and rats) annually in this country have ranged from 17 to 70 million[23]; other sources report that about 35 to 50 million laboratory animals are bred in the United States every year with some 25 to 30 million actually used.[24]

In 1976, again the Animal Welfare Act was amended, but this time the changes centered on transportation and animal fighting ventures. The most relevant (to research) of these amendments reordered the priorities of the act's objectives. The objectives, now listed in reverse order from previous versions of the law, are "(1) to insure that animals intended for use in research facilities or for exhibition purposes or for use as pets are provided humane care and treatment; (2) to assure the humane treatment of animals during transportation in commerce; and (3) to protect the owners of animals from the theft of their animals by preventing the sale or use of animals which have been stolen." The 1976 amendments also increased the Secretary's options for enforcement and penalties.

In 1985, after almost 20 years of tumultuous activity and the exchange of accusations among scientists and animal protection groups, NIH opened the current round of change by revising its policy governing research in facilities funded by the PHS.[25] Soon thereafter (on November 20, 1985), Congress passed the Health Research Extension Act, which gave legislative force to PHS policy.[26] On December 23, 1985, shortly after the passage of the Health Research Extension Act, Congress amended the Animal Welfare Act for the third time.

There are major differences between the Laboratory Animal Welfare Act of 1966 and both the amended Animal Welfare Act and the new PHS Policy. The Animal Welfare Act was initiated in 1966 primarily to prevent animal theft and the sale or use of stolen animals; the principal intent of the new regulations is the assurance of humane care and treatment of laboratory animals. This is an objective that has taken 2 decades to develop.

The Amended Animal Welfare Act

The 1985 amendments change the Animal Welfare Act in several important ways (see Appendix A for the complete text of the amended Animal Welfare Act):

1. Each research facility covered by the Animal Welfare Act, including federal facilities, is now required to establish an Institutional Animal Committee, composed of not fewer than three members, possessing sufficient ability to assess animal care, treatment, and practices in experimental research with animals. At least one of these members should be a veterinarian, and one should be unaffiliated with the institution and is intended to provide representation of community interests. Any member of the committee "willfully" revealing confidential information is subject to a fine of up to $10,000 and 3 years imprisonment. The committee must inspect all animal study areas at least twice a year, to assess the condition of the animals and to review practices involving pain to the animals.
2. Provision must be made for the exercising of dogs.
3. Provision must be made for a physical environment adequate to promote the psychological well-being of primates. (At the present time, guidelines for this provision have not been completed.)
4. An animal should not be used in more than one major operative experiment from which it is allowed to recover, except in cases of scientific necessity or by determination of the Secretary of Agriculture.
5. An information service is to be established at the National Agricultural Library (NAL). In cooperation with the National Library of Medicine, the service is to provide information that could prevent the unintended duplication of animal experiments, reduce or replace animal use, minimize animal pain or distress, and aid in the training of personnel involved with animals.
6. Training for personnel who handle animals must be provided.

Because of the changes in the Animal Welfare Act enacted in 1970, most research facilities in the United States that use regulated species must comply with these requirements. (As noted, facilities that do not buy animals in interstate commerce or receive federal funds are still exempt.) By the end of 1986, a total of 1,241 research facilities had registered with APHIS as using animals covered by the act.[27] The provisions of the Animal Welfare Act are applied through the PHS Policy, which requires that all institutions conform not only with the policy, but also with the Animal Welfare Act.

The New PHS Policy

The new PHS Policy is applicable to all federal awardee institutions (as before) and to all federal agencies conducting research involving animals (a new provision). The policy was developed in conjunction with an extensive 2-year working group at NIH, consultations with experts in many disciplines, as well as site visits to more than 15 institutions. In addition, hearings were held by NIH at various regional centers, and almost 400 written and oral comments on the proposed policy were studied and considered. To provide a basis for understanding the discussion in the remainder of this chapter, we refer the reader to the text of sections of the Health Research Extension Act of 1985 that apply to animal research, included in this volume as Appendix B.

There are three major changes in the new policy with which each federal awardee institution must comply. Each research entity must (1) designate clear lines of authority and responsibility for those involved in the care and use of laboratory animals. Two named officials must be specified, one the institutional official responsible for administering the program and the other the veterinarian who will participate in the program; (2) establish an Institutional Animal Care and Use Committee, which is to be involved in all aspects of animal care and use, and is to review and approve components of all proposals to federal agencies regarding the care and use of animals; and (3) file a written Animal Welfare Assurance with the NIH.

The first requirement gives authority, for the first time, to the veterinarian in the institution. Previously, most veterinarians involved with laboratory animal medicine in research institutions had little to no power or authority designated to them by the administration. A veterinarian's intervention in an experiment or an investigator's seeking advice from such a resource occurred infrequently. Since the new policy requires compliance with the Animal Welfare Act, the veterinarian must be consulted if the procedure is likely to cause pain to the animal.

The second change, the requirement for an Institutional Animal Care and Use Committee (IACUC), specifies that this committee be composed of not less than five members, including at least one veterinarian, one scientist experienced in animal research, one nonscientist, and one individual not associated with the organization (the same person may fulfill more than one requirement).[28] This provision enhances the veterinarian's authority by assuring that he or she will have input in ongoing activities at every level of animal use. The nonaffiliated member of the committee is intended to bring objectivity into the review of proposals and can represent the interests of the community concerning the humane treatment of animals in research. The policy charges this committee with reviewing research proposals initiated at the institution, in order to assess the humaneness of animal care and treatment during the proposed research. Unless the IACUC grants approval, federal agencies will not consider the application for funding. In the review of proposals, the IACUC shall determine that the proposed research conforms with the institution's assurance (see below) and meets the following requirements:

1. Procedures with animals will avoid or minimize discomfort, distress, and pain to the animals, consistent with sound research design.
2. Procedures that may cause more than momentary or slight pain or distress to the animals will be performed with appropriate sedation, analgesia, or anesthesia, unless the procedure is justified for scientific reasons in writing by the investigator.
3. Animals that would otherwise experience severe or chronic pain or distress that cannot be relieved will be painlessly sacrificed at the end of the procedure or, if appropriate, during the procedure.
4. The living conditions of animals will be appropriate for their species and will contribute to their health and comfort. The housing, feeding, and nonmedical care of the animals will be directed by a veterinarian or other scientist trained and experienced in the proper care, handling, and use of the species being maintained or studied.
5. Medical care for animals will be available and provided as necessary by a qualified veterinarian.
6. Personnel conducting procedures on the species being maintained or studied will be appropriately qualified and trained in those procedures.
7. Methods of euthanasia will be consistent with the recommendations of the American Veterinary Medical Association Panel on Euthanasia, unless a deviation is justified for scientific reasons in writing by the investigator.[29]

The committee also has the responsibility to assure that investigators maintain compliance with the PHS Policy and the Animal Welfare Act. The committee is to inspect the facilities on a semiannual basis and file reports with the Office for Protection from Research Risks (OPRR) at the NIH. Thus, the IACUC is in effect a local ethical committee that can give on-site review and approval of all aspects of animal use at the institution.

The final requirement, the filing of an "Animal Welfare Assurance" by the institution to the OPRR, is perhaps the most significant of the changes. No activity involving animals may be conducted or supported by the PHS until the institution conducting the activity has provided the OPRR with a written assurance acceptable to the PHS, setting forth compliance policy. This assurance describes the adequacy of the institution's proposed program for the care and use of animals in PHS-conducted or supported activities. The institution must provide details of its program for the care and use of animals using the guide as the basis for developing and implementing an institutional program for all activities involving animals. Approval of the assurance will be for a specified time (no longer than 5 years), after which a new assurance must be submitted for approval. Approval of the assurance by the OPRR is tantamount to giving the institution a license to conduct research on its premises. The PHS Policy designates the head of the institution and the IACUC as having the responsibility to ensure that all investigators are in compliance with that assurance.

The NIH has been concerned with the question of "assurance" for more than 20 years. The definition of the concept and the process whereby it could be implemented and enforced also took almost 2 decades. Table 2.1 gives details of the stages in the development of the concept of assurance by the PHS.

In addition to requiring adherence to the provisions of the Animal Welfare Act and the development of assurance by the institution, based on the guide, the new PHS Policy also requires that institutions comply with all other applicable provisions of federal statutes and regulations relating to animals, and be guided by the United States Government Principles for the Utilization and Care of Vertebrate Animals Used in Testing, Research, and Training.

Thus, the two newly legislated acts together affect a much larger component of the research community than previously, as was evidently the intent of their Congressional sponsors. According to the Office of Technology Assessment, "This provision for an institutional animal committee [by the amended Animal Welfare Act of 1985] taken in concert with similar provisions in the Health Research Extension Act [and the new PHS Policy] brings the overwhelming majority of animal researchers in the United States under the oversight of a structured local review committee."[30] However, as the act was not rewritten to cover rats and mice, the current regulations still do not affect a large proportion of animals used in research and testing.

It is significant that animal pain and distress during experimentation were directly confronted for the first time in these new regulations. This was accomplished by giving the institutional committees a mandate, under both the PHS Policy and the amended Animal Welfare Act, to consider pain and distress in their review of research practices, and by giving the facility's veterinarian the authority to bring his or her professional expertise to bear. Most scientists are aware that a distressed animal is not an appropriate experimental subject, but defining and recognizing pain and distress in different animal species and under different conditions are extremely difficult to do. We will return to this subject in Chapter 5, where we show that the American Veterinary Medical Association has

TABLE 2.1. PHS Policy and development of Animal Welfare Assurance.[a,b]

1963 **Guide for Laboratory Animal Facilities and Care**
—No assurance required
—Guide is recommended

1971 **NIH Policy on Care and Treatment of Laboratory Animals**
—Two-paragraph standard language assurance
—AAALAC accreditation *or* evaluation committee with veterinarian
—If no AAALAC accreditation, then periodic committee inspection of animal facilities
—Adherence to the guide, NIH guidelines, and AWA

1973 **DHEW Policy on Animal Welfare**
—Two- or three-paragraph standard language assurance, including
—name of veterinarian and
—list of institutional components
—AAALAC accreditation *or* three-member Institutional Evaluation Committee
—If no AAALAC accreditation, then continuing committee evaluation of animal care, use, and treatment
—Adherence to the guide, DHEW principles, and AWA

1979 **PHS Policy on Humane Care and Use of Laboratory Animals**
—Assurance in same format as 1973 Policy required prior to award; 5-year duration
—AAALAC accreditation *or* five-member institutional committee
—If no AAALAC accreditation, then annual committee review of animal procedures and facilities
—Adherence to the guide, PHS principles, and AWA
—Annual report to OPRR if nonconformance cited
—Institutional review of proposals encouraged

1985 **PHS Policy on Humane Care and Use of Laboratory Animals by Awardee Institutions**
Awardee institutions must give written assurance, acceptable to the OPRR, that they will comply with the new policy before conducting activities with animals supported with PHS funds. This assurance must include information pertaining to the following requirements:
—five-member Institutional Animal Care and Use Committee (IACUC), which must
—review and approve all PHS proposals,
—conduct annual inspections of program and facilities, and
—submit annual reports to the OPRR
—Recordkeeping by the institution
—Institutional status: AAALAC approval *or* institutional evaluation
—Lines of institutional authority and responsibility, including
—name of veterinarian and
—name of institutional official
—Adherence to guide, AWA, and U.S. Government Principles
—Use of guide as basis for institutional program involving animals
—Description of health program for all personnel working with animals

1986 **PHS Policy on Humane Care and Use of Laboratory Animals**
Assurance as in 1985, with the following additions and changes:
—PHS Policy also applicable to research that the PHS conducts (as well as awardee institutions), and written assurance must now be made by these agencies

continued

TABLE 2.1. *(continued)*

–Description must be included of institutional program of education and training available to personnel working with animals
–IACUC now appointed by chief executive officer of institution
–*Semi*annual review now required of facilities and program by IACUC
–IACUC, through institutional official, to be responsible for reporting requirements
–Minority views must be included in annual reports of IACUC to OPRR

[a]We thank Dr. John Miller of the Office for Protection from Research Risks of the National Institutes of Health for providing information included in this table.
[b]Table 2.1 shows the development of the assurance process by the National Institutes of Health (NIH) and the Public Health Service (PHS). *DHEW* indicates the former Department of Health, Education, and Welfare (now the Department of Health and Human Services); *AWA* refers to the Animal Welfare Act; *AAALAC* indicates the American Association for Accreditation of Laboratory Animal Care; the "guide" refers to the *Guide for the Care and Use of Laboratory Animals*; "U.S. Government Principles" refer to the *U.S. Government Principles for the Utilization and Care of Vertebrate Animals Used in Testing, Research, and Training*; and "OPRR" refers to the Office for Protection from Research Risks of the NIH.

been concerned with this problem since about 1975, but that overall it has received little attention from other scientists.

It is also noteworthy that both the Health Research Extension Act and the 1985 AWA contain recognition of the importance of alternative methods. As noted in a report issued by the Office of Technology Assessment, 1985 was the first year in which the concept of alternative methods was explicitly described in federal law.[31] Actually, the idea of alternatives appeared in *three* federal laws that year: The other was the Health Professions Educational Assistance Act of 1985 (Public Law 99–129), which mentions the inclusion of alternatives to the use of animals in the development of curricula for veterinary students.

Although the Animal Welfare Act of 1966 was the first federal law to begin to address the needs of laboratory animals in the United States, it was an indirect and cautious effort. The act in its original form, even as amended in the 1970s, was so broad as to be ambiguous. It did not recognize the problem of pain and distress during experimental procedures, nor did it consider the conditions under which experiments would be conducted. Moreover, the USDA appeared reluctant to assume the responsibility for enforcing the act, nor was it funded adequately to do so. The very large number of institutions, and the even larger number of investigators, in the United States (as compared with those in the United Kingdom or Scandinavian countries, for instance) makes most unwieldy a federal inspection system such as the one contained in the original act. Hence, implementation and enforcement of the act seemed necessarily inefficient and ineffective.

The new PHS Policy, in coordination with the latest version of the Animal Welfare Act, is, in comparison, a clearly defined set of regulations with legislative force behind it. The policy is intended to be broad without being ambiguous, specific without being restrictive. Under an "Animal Welfare Assurance," it also, for the first time, places on both the institution and the investigator the responsibility for the humane treatment of laboratory animals and for the ethical conditions under which the experiments are carried out.

Whether or not the new American approach to regulation of animal research will be given a chance to work depends to a great extent on the attitudes of the researchers themselves. It is therefore of great practical importance, as well as a subject of historical significance, to attempt to describe and document current attitudes and to trace their development and change over the past 2 decades.

Notes

(The notation *[AB]* after a reference indicates that it is included in the Annotated Bibliography following Chapter 7.)

1 Ritvo H (1984) Plus ça change: Antivivisection then and now. *Sci Tech Human Values 19*,2:57 [AB].

2 Benison S (1970) In defense of medical research. *Harv Med Alumni Bull 44*,3:16 [AB]. It was not until nearly 1960 that humane societies in the United States began to express concerns about the use of animals in experimentation. Today most, if not all, humane societies have expressed reservations about the use of animals in research. They hold a wide spectrum of positions on this question. Some want to reduce the number of animals used and to limit the types of experiments to be carried out with animals; some are concerned primarily with conditions of housing animals. Still others are most critical of the use of domestic animals for research.

3 Ibid, p 17.

4 Ibid, pp 17–23; Corner G (1964) *A history of the Rockefeller Institute.* New York: Rockefeller University Press, pp 83–87.

5 *Guide for laboratory animal facilities and care.* Department of Health, Education, and Welfare, Publication 1024 (1963). Prepared by the Animal Facilities Standards Committee of the Animal Care Panel and adopted by the Institute for Laboratory Animal Resources (ILAR) on January 28, 1963. The 1963 guide was subsequently adopted as policy by the PHS. Successive revisions have been published under the title, *Guide for the care and use of laboratory animals,* hereinafter referred to as the guide.

6 Levine has pointed out that it was also in 1966 that the first public policy designed to safeguard the rights and welfare of *human* research subjects was instituted in the United States. The federal policy issued that year by the Surgeon General required peer review of projects involving human subjects in research supported by the PHS (Levine RJ [1988] Protection of human subjects of biomedical research in the United States: A contrast with recent experience in the United Kingdom. In: Callahan D, Dunstan GR [eds] *Biomedical ethics: An Anglo-American dialogue.* Annals of The New York Academy of Sciences, vol 530. New York: The New York Academy of Sciences, pp 133–134).

[7] The Animal Welfare Act. Title 7, US Code, Chapter 54, para 2131–2157. Public Law 89–544 (1966) and amendments: Public Law 91–579 (1970), Public Law 94–279 (1976), and Public Law 99–198 (1985). See Appendix A for the complete text of the law in its current form.

The Public Health Service Act. Title 42 US Code, Chapter 201 *et seq*, Public Law 78–184 (1944), amended by the Health Research Extension Act of 1985, Public Law 99–158 (1985): NIH Reauthorization, Chap 289d) sec 495a). See Appendix B for the relevant text of the Health Research Extension Act.

[8] United States Constitution. Article I, Sec 8; Amendment X.

[9] Sechzer JA, Grodsky PB (1978) History and current status of antivivisection, humane, and animal rights groups in the United States. Program of the 86th Annual Meeting of the American Psychological Association; French RD (1975) *Antivivisection and medical science in Victorian society*. Princeton: Princeton University Press, pp 392–412; Griffin A, Sechzer JA (1983) Mandatory versus voluntary regulation of biomedical research. In: Sechzer JA (ed) *The role of animals in biomedical research*. Annals of the New York Academy of Sciences, vol 406. New York: The New York Academy of Sciences, pp 187–200 [AB].

[10] The most notorious cases have occurred in the 1980s. In 1981, 18 monkeys were seized from the laboratory of Dr. Edward Taub by police in Silver Spring, Maryland, after a tip-off by animal protectionists that animals there were being abused. Taub was convicted by a Maryland court on six counts of failure to provide proper veterinary care, and his NIH funding was rescinded. The convictions were later overturned on a technicality, but Taub's NIH grant was not restored. See Monkeys transferred after abuse charges (September 23, 1981) *The New York Times*; Holden C (1981) Police seize primates at NIH-funded lab. *Science 214*:32–33 [AB]; Holden C (1981) Scientist convicted for monkey neglect. *Science 214*:1218–1220 [AB]; Sechzer JA (ed) *The role of animals in biomedical research*. Annals of the New York Academy of Sciences, vol 406. New York: The New York Academy of Sciences, p 150 [AB]; Fox JL (1984) Animal rights bill defeated in California. *Science 224*:1414 [AB].

In 1984 a set of videotapes was stolen from the University of Pennsylvania Medical School's head trauma laboratory by a group calling itself the Animal Liberation Front. Approximately 70 hours of videotape were condensed to about 35 minutes and were subsequently displayed across the United States as well as in several European countries. Researchers were seen carrying out surgical procedures on baboons with inadequate sterile technique, smoking during surgery, and in general showing a lack of sensitivity toward animals that had undergone severe head trauma. Research funds were withdrawn by the NIH, and for failure to comply with the Animal Welfare Act, the university was fined $4,000. See Fox JL (1984) Lab break-in stirs animal welfare debate. *Science 224*:1319–1320 [AB]; Culliton BJ (1985) HHS halts animal experiment. *Science 229*:447–448 [AB].

In another 1984 episode, animal activists broke into the laboratories of the City of Hope Medical Center in Duarte, California. The group removed animals as well as documents that substantiated serious violations of regulations. An NIH investigation resulted in the suspension of 31 research projects, and the center was fined $11,000 by the USDA.

Another action occurred in January, 1986, at Columbia University's College of Physicians and Surgeons in New York City. Photographs and other evidence of violations in the animal facilities were produced after a raid by the Animal Liberation Front. Shortly after the implementation of the PHS Policy, the NIH responded with an unannounced inspection, resulting in suspension of all research on animals above the level of rodents. The university was also fined. The conditions in the facilities were corrected and reinspected before funding was restored later in the year. (We thank the National Association for Biomedical Research for information describing these latter two events provided to us in various issues of *Update*.)

[11] Caplan AL (1983) Beastly Conduct: Ethical issues in animal experimentation. In: Sechzer JA (ed) *The role of animals in biomedical research*. Annals of the New York Academy of Sciences, vol 406. New York: The New York Academy of Sciences, pp 159–169 [AB].

[12] Erbe NA (1966) *Anti-cruelty laws and scientific use of animals in the United States*. London: Research Defense Society, pp 1–12. This booklet describes the anticruelty laws in the United States through 1966.

[13] Hearings were held in March and May 1966 to consider dozens of related bills, one of which (H.R. 13881) was enacted as the Animal Welfare Act. Additional hearings were held on some of the other bills. However, relatively little was written about them in the literature. See Chapter 4 for a comparison of the amount of discussion generated by the 1966 act compared to that of the 1980s.

[14] *Hearings before the Subcommittee on Science, Research, and Technology of the Committee on Science and Technology*. (1981) U.S. House of Representatives, 97th Congress, 1st Session (No 68).

[15] One such group was the American Association for Accreditation of Laboratory Animal Care (AAALAC), organized in 1965 for the purpose of conducting a voluntary program to accredit institutions having laboratory animal facilities. AAALAC's primary function is to improve animal care programs by means of an accreditation process that involves conducting site visits to institutions, evaluating site visit reports, and making recommendations to its Board of Trustees concerning proposed accreditation status of the institutions visited. This program encourages very high standards for the care and use of laboratory animals. These standards include providing appropriate veterinary care, becoming aware of how environmental factors might affect animal research, and assuring the safety of personnel who work with the animals. Since one of the AAALAC principles includes adherence to the PHS guide, NIH has recognized AAALAC accreditation as a means of demonstrating compliance with PHS Policy.

AAALAC may have been organized initially in 1965 to avoid or defer legislative regulation of animal research, but it continues to play an important role in the escalating concern for the use of animals in experimentation. However, the AAALAC program has not yet received wide acceptance from research facilities. As of December 31, 1987, only about 700 out of approximately 1,600 institutions using regulated species had sought AAALAC accreditation, with 521 organizations accredited (American Association for the Accreditation of Laboratory Animal Care [1988] *Information brochure*. Bethesda, MD).

[16] United States Department of Agriculture, Animal and Plant Health Inspection Service (1983; 1987) *Animal welfare enforcement, FY 1982* and *FY 1986: Report of the Secretary of Agriculture to the President of the Senate and the Speaker of the House of Representatives.*

[17] National Research Council (1988) *Use of laboratory animals in biomedical and behavioral research.* Washington, DC: National Academy Press.

[18] Federal regulation of animal use. (1986) In: U.S. Congress, Office of Technology Assessment, *Alternatives to animal use in research, testing, and education.* Washington. DC: U.S. Government Printing Office, OTA-BA–273, p 278.

[19] Idem. The regulations in their current form are published in *Code of Federal Regulations (CFR)* (1985) Title 9, Chap 1. Animal and Plant Health Inspection Service, USDA. Subchapter A—Animal Welfare. Washington, DC: U.S. Government Printing Office: 1986–180–916:40234, Washington–1–85, Sec 2,28, p 1.

[20] Dr. Dale F. Schwindaman, United States Department of Agriculture, personal communication to J.A.S., 1983; Orlans FB (1981) Scientists Center for Animal Welfare, Statement for the record. In: *Hearings before the Subcommittee on Science, Research, and Technology of the Committee on Science and Technology.* U.S. House of Representatives, 97th Congress, 1st session (No 68), pp 587–590.

[21] *Code of Federal Regulations* (1985) Op cit [note 19], p 9.

[22] United States Department of Agriculture, Animal and Plant Health Inspection Service (USDA-APHIS) (1987) *Animal welfare enforcement, FY 1986: Report of the Secretary of Agriculture to the President of the Senate and the Speaker of the House of Representatives.*

[23] Orlans BF (1981) What institutional animal research committees can do to improve humane care. *Lab Animal* (UK) *13*:24–29; Rowan A (1984) *Of mice, models, & men: A critical evaluation of animal research.* Albany, NY: State University of New York Press. An estimate of 17 to 22 million was given in National Research Council (1988) Op cit [note 17].

[24] Patterns of animal use. (1986) In: U.S. Congress, Office of Technology Assessment, op cit [note 18], pp 43–68.

[25] Office for Protection from Research Risks (OPRR) (June 25, 1985) *Public Health Service policy on humane care and use of laboratory animals by awardee institutions.* Bethesda, MD. The *Guide for the care and use of laboratory animals* also underwent revision in 1985 to reflect the new policy. See *Guide for the care and use of laboratory animals.* Public Health Service, National Institutes of Health (NIH), Publication No. 86–23, revised, 1985.

[26] The Health Research Extension Act gave PHS policy a specific basis in the law. In doing so, it required changes to be made in the existing PHS policy. The policy was extended to apply to all federal agencies conducting animal research, which had previously been exempt. To accommodate these changes, NIH again revised the policy. On September 5, 1986, the current revision was issued as *Public Health Service policy on humane care and use of laboratory animals* (hereinafter referred to as PHS Policy). See Office for Protection from Research Risks (OPRR) (September 5, 1986) *Public Health Service policy on the humane care and use of laboratory animals.* Bethesda, MD; see Appendix B for the text of relevant portions of the Health Research Extension Act of 1985.

[27] USDA-APHIS (1987) Op cit [note 22].

[28] In requiring a minimum of five members for the Institutional Animal Care and Use Committee (IACUC), the new PHS Policy exceeds the requirements of the 1985 Animal Welfare Act amendments and the Health Research Extension Act, both of which require the committee to have a minimum of only three members.

[29] OPRR (September 5, 1986) Op cit [note 26], pp 7–8.
[30] Federal regulation of animal use (1986) Op cit [note 18].
[31] Ibid.

CHAPTER 3

Literature Survey: Procedures and Methods

The comments and conclusions of the following two chapters are based on a survey of the American scientific literature conducted by the authors. As noted in the Introduction, this survey was originally intended merely to provide information to aid in the development of a set of interdisciplinary guidelines for the use of animals in experimentation. By analyzing changes over time in both the quantity and the content of items located in the literature, however, we have made further use of the survey data.

In this chapter we describe our methods for generating the data used in the analyses in Chapters 4 and 5. To ensure reliability, we used carefully defined criteria to determine which articles or other items from the scientific literature would be counted, and we employed systematic procedures in conducting the search.

Criteria for Selection of Relevant Items

First, we included only those items published from 1966 through the end of 1985. As already indicated, these years demarcate a period of transition in federal policies concerning the care and use of laboratory animals.

Second, we limited the search to journals and books published in the United States. This might be questioned on the grounds that American scientists regularly read and contribute to English-language international journals in their fields, such as *Pain* (published in Amsterdam) and the *International Journal of Neuroscience* (published in England). Such journals probably do reflect and influence the thinking of the American scientific community, but they also reflect other concerns, not necessarily shared in the United States. For this reason, and because we needed to limit the search to manageable proportions, we omitted all foreign publications. However, we included several books published in the United States that were also published elsewhere.

Third, we included only items that could reasonably be considered to have been published by the scientific community or for a professional scientific audience. This criterion meant that we omitted the sizable literature emanating from humane societies, animal welfare organizations, and animal rights groups. Peter Singer's *Animal Liberation*, for instance, was not included, even though it has had an enormous impact in mobilizing the movement that led indirectly to the writing of this book. On the other hand, a review of *Animal Liberation* was included because it appeared in *Science*, a magazine published expressly for scientists.[1]

Decisions in applying this criterion were not always clear-cut, as in the case of two books written by an antivivisectionist who is also a physician. These books were published by an animal welfare organization founded by the author, but because of his medical credentials, we included the books anyway.[2] In another instance, we omitted numerous articles and several books by a biochemist because the author wrote as an official of the Humane Society of the United States (HSUS). These books are nevertheless valuable resources for scientists interested in reducing the number of animals used in research, refining experimental techniques to reduce animal suffering, and replacing animals with in vitro methods where possible.[3] The same author wrote a third book after he moved from HSUS to take an academic post in the research community; the latter book has been included.[4]

A special problem was presented by *The International Journal for the Study of Animal Problems* (IJSAP), published by the HSUS from 1980 through 1983. This was an excellent publication directed to a scientific audience as well as to animal welfare advocates. The hundreds of articles and other items that appeared in this publication over its brief life would have qualified for inclusion in this study by all other criteria, but in the end we thought that doing so would inflate the results in a misleading way. The IJSAP is highly recommended to readers interested in pursuing issues raised in this book, but it is more realistically considered a product of the animal protection movement than of the scientific community.

The professional scientific criterion also forced us to omit works by philosophers and historians, unless their writings appeared in the scientific literature. For instance, we did not include the fascinating historical studies by Richard D. French, James Turner, or Coral Lansbury.[5] When articles by French and philosopher/activist Peter Singer appeared in the *Encyclopedia of Bioethics*, however, we included them,[6] as we did reviews of the French and Turner books published in the *Bulletin of the History of Medicine*.[7]

The final criterion is concerned with the content matter of the items, and this was the most difficult to apply consistently. Our guiding principle was that, to be considered relevant, an item should in some way address the controversy over the use of animals in research. All articles that discussed the welfare of laboratory animals, the ethics of animal experimentation, or regulatory measures regarding laboratory animals were judged relevant. Also included were articles that defended the use of animals in research by surveying medical progress made through animal experimentation, even if ethical or welfare issues were not explicitly mentioned. Editorials, book reviews, letters, columns, committee reports, and other miscellaneous items (a cartoon, some reprinted excerpts from articles in other journals, guidelines, etc) were also included, as long as they dealt with relevant issues. Even news briefs concerning the progress of regulatory legislation were included, although they sometimes consisted of only a few lines each. The rationale was that each item, no matter how small, indicates an interest in the subject and should be counted as a measure of this interest. In Chapter 4 we take into account the unequal significance of short news briefs or letters compared to longer, weightier works.

Procedures for Conducting the Literature Search

Items included in the analysis were located in three ways: by a computerized literature search; by an exhaustive manual search of five selected journals; and by following leads in references, footnotes, and bibliographies.

The major databases used in the computer search were Index Medicus, Excerpta Medica, Biological Abstracts, Psychological Abstracts, and the Social Sciences Citation Index. Together these databases index thousands of journals in the biological, medical, and social sciences. Each item cited in the computer search was checked for the criteria listed above, and those that did not fulfill all four criteria were discarded. A complete list of all journals containing items that fulfilled our criteria is given in Appendix E.

To test the comprehensiveness of the computer search, we also searched five journals manually. The five journals were selected from different fields of science, and all were indexed by at least one (usually several) of the databases. The journals searched manually were the *New England Journal of Medicine*, *Physiologist*, *Science*, *Journal of the American Veterinary Medical Association*, and *American Psychologist*.

We were surprised by the poor showing made by the computer search in this test: Of 484 items found manually, the computer found only 62, or about 13% (see Table 3.1). Of course, one would not expect electronic databases to index brief, untitled, anonymous news items; the large number of relevant news briefs in the *Journal of the American Veterinary Medical Association* accounts for that publication's extremely low percentage of computer-found items. Nevertheless, the computer also missed major feature articles. An important, lengthy article defending animal research[8] and a much-cited discussion of the ethical issues raised by some classic 19th-century experiments,[9] both in *American Psychologist*, were ignored by the electronic databases. Also passed over was a feature report on the activities of animal welfare activists and legislative developments

TABLE 3.1. Literature search for items related to research animal welfare: Results of a computerized search compared with a search done manually, for five selected scientific journals, 1966–1985.

	Total number of relevant items found, 1966–1985		
	Computer search	Manual search	Items found by computer, as a percentage of all items
New England Journal of Medicine	9	17	52.9%
Physiologist	6	64	9.4%
Science	25	89	28.1%
Journal of the American Veterinary Medical Association	17	298	5.7%
American Psychologist	5	16	31.3%
Totals	62	484	12.8%

published in the *New England Journal of Medicine.*[10] An editorial in *Science* expressing alarm at the threat to animal research posed by animal welfare activists,[11] and news features in the same magazine about developments in the controversy[12] are a few more examples of articles we would have missed had we relied solely on the computer search. In the *Physiologist*, the computer search located "APS statements on NIH Guide revisions,"[13] but missed another item with virtually the same title that had appeared a few months earlier.[14]

The *New England Journal of Medicine* was the only journal to have more than half of its relevant items located in the computer search. It fared better than the others, partly because all of these items were letters, articles, or book reviews—no news briefs or other miscellaneous tidbits that were unlikely to be indexed. If articles and letters alone are tabulated for the *American Psychologist* (there were no relevant book reviews), the latter's percentage of computer-found items rises to 41.7%. Even this higher rate is a sobering discovery for those who put complete faith in computerized literature searches. Such results are disturbingly capricious and obviously far from exhaustive.

Nevertheless, we have confidence in the validity of the results reported in the following chapters. This confidence is warranted for several reasons: First, there is no reason to believe that there is a systematic bias in the items indexed by the electronic services. Other than the latter's tendency to omit smaller items, the differences between articles missed and those indexed are probably not significant for our purposes.

Second, the number of items used in the analysis is large enough to form a basis for reasonably accurate conclusions, and the items represent a wide range of journals, in spite of the fact that the data are not exhaustive. From all sources, we located a total of 690 items that fulfilled our criteria for relevance; 659 of these were journal items published in 79 different journals (see Appendix E).

Third, we confirmed our results by performing separate computations using only the 484 items found in our manual search of the five selected journals. The trends discovered in these data, which we knew to be exhaustive for the five journals, were quite similar to the results for the total body of data. These trends, and an analysis of the small differences that did appear between the two sets of data, are presented in Chapter 4.

Papers published in books rather than journals were handled variously, depending on the nature of the book. If the book as a whole dealt primarily with the issues of laboratory animal welfare, the ethics of animal research, or regulatory issues, the book was counted once (ie, as one item), and the papers were not counted individually. Examples of books of this kind include a volume of papers presented at a 1981 conference on animal welfare sponsored by the Scientists Center for Animal Welfare and the proceedings of a conference the following year sponsored by the New York Academy of Sciences.[15] There are 13 books handled in this way included in the total of 690 items.

If, on the other hand, a book contained one or a few relevant papers but its central focus was *not* relevant for our purposes, then the papers were counted individually, and the book was not counted as a separate item. There are 18 items of this type included in the total. One example is an intriguing paper by Colin Beer that addresses the philosophical problem of other minds as it applies to the ethics of animal experimentation.[16] The remaining papers in this volume deal with ethical problems in the treatment of human patients and the general implications of neurology and biology for ethics, with no mention of animal research. This book, therefore, was not counted as a separate item.

We made an effort to locate and read each of the 690 items, and we succeeded in over 90% of the cases. Items that were not read tended to be ones that were unavailable in major research libraries in the New York area, and thus are probably also the items with the smallest readership in the scientific community. For every item that we read, we prepared a brief summary and coded its contents for topics, tone, and themes. These summaries and coded categories were used for the qualitative analysis presented in Chapter 5.

In doing the thematic coding, we departed from the procedure for handling articles in books described above. Since it was impossible to code a collection of papers by different authors for dominant ideas or themes as though the book formed a coherent whole, we coded each paper separately, regardless of whether it had been counted individually in the quantitative analysis. For this reason, the total number of items included in the qualitative analysis in Chapter 5 (717) is larger than the total number used in the quantitative analysis in Chapter 4 (690).

The annotated bibliography consists of selections made from all the items. We thought it most helpful to make separate bibliographic entries for *both* full-length books *and* individual papers published in collected volumes, since here the numerical count is irrelevant. In addition, we added some articles from *Lab Animal*, a trade journal published in New York for veterinarians and animal facility technicians. Trade journals were not included in the quantitative data, but these articles are of sufficient interest to justify including them in the annotated bibliography.

Notes

(The notation *[AB]* after a reference indicates that it is included in the Annotated Bibliography following Chapter 7.)

1 Singer P (1975) *Animal liberation.* New York: New York Review/Random House; Hull D (1976) The rights of animals. Review of Peter Singer, *Animal liberation. Science 192*:679–680 [AB].

2 Pratt D (1976) *Painful experiments on animals.* New York: Argus Archives [AB]; Pratt D (1980) *Alternatives to pain in experiments on animals.* New York: Argus Archives [AB].

3 Rowan AN (1979) *Alternatives to laboratory animals.* Washington, DC: Humane Society of the United States; Rowan AN, Stratman CJ (eds) (1980) *The use of alternatives in drug research.* Baltimore, MD: University Park Press (also published in London by Macmillan).

4 Rowan AN (1984) *Of mice, models, & men: A critical evaluation of animal research.* Albany, NY: State U of New York Press [AB].

5 French RD (1975) *Antivivisection and medical science in Victorian society.* Princeton, NJ: Princeton U Press; Turner J (1980) *Animals, pain and humanity in the Victorian mind: Reckoning with the beast.* Baltimore, MD: Johns Hopkins U Press; Lansbury C (1985) *The old brown dog: Women, workers, and vivisection in Edwardian England.* Madison, WI: The U of Wisconsin Press. [See Introduction, notes 3–5.] Although we did not include these works in the tabulations analyzed in Chapters 4 and 5, we did make extensive use of them (particularly French) as source materials for Chapter 1.

6 French RD (1978) Historical aspects [of animal experimentation]. In: Reich WT, et al (eds) *Encyclopedia of bioethics* vol 1. New York: Free Press, pp 75–79 [AB]; Singer P (1978) Philosophical perspectives [on animal experimentation]. In: Reich WT et al (eds) *Encyclopedia of Bioethics* vol 1. New York: Free Press, pp 79–83 [AB].

7 Mazumdar PMH (1978) Antivivisection and medical science in Victorian society [book review]. *Bull Hist Med 52*,4:601–602 [AB]; Loew FM (1982) Animal experimentation [book review]. *Bull Hist Med 56*,1:123–126 [AB].

8 Miller NE (1985) The value of behavioral research on animals. *Am Psychol 40*,4:423–440 [AB].

9 Gallistel CR (1981) Bell, Magendie and the proposals to restrict the use of animals in neurobehavioral research. *Am Psychol 36*,4:357–360 [AB].

[10] Inglehart JK (1985) Health policy report: The use of animals in research. *N Engl J Med 313*,6:395–400 [AB].

[11] Hess CE (1984) Freedom of enquiry: An endangered species. *Science 225*:273 [AB].

[12] Fox JL (1984) Rifkin takes aim at USDA animal research. *Science 226*:321 [AB]; Culliton BJ (1985) HHS halts animal experiment. *Science 229*:447–448 [AB].

[13] American Physiological Society (1983) APS statements on NIH *Guide* revisions. *Physiol 26*,5:273–274 [AB].

[14] American Physiological Society (1983) APS statement on NIH *Guide* revisions. *Physiol 26*,3:113–114 [AB].

[15] Dodds JW, Orlans FB (eds) (1982) *Scientific perspectives on animal welfare.* New York: Academic Press [AB]; Sechzer JA (ed) (1983) *The role of animals in biomedical research.* Annals of the New York Academy of Sciences, vol 406. New York: The New York Academy of Sciences [AB].

[16] Beer C (1983) Motives and metaphors in considerations of animal nature. In: Pfaff DW (ed) *Ethical questions in brain and behavior: Problems and opportunities.* New York: Springer-Verlag, pp 125–140 [AB].

CHAPTER 4

Quantitative Analysis: Measures of Interest

Literature on the moral status of animals has proliferated in recent years: One estimate has it that more has been written on this topic in the past dozen years than in the previous 3,000.[1] It does not necessarily follow, however, that the growing public preoccupation with animal rights has any counterpart in the scientific literature. In fact, researchers have frequently been taken to task for ignoring animal welfare issues. Scientists themselves have sometimes concurred, as in 1977 when the Federation of American Scientists (FAS) published a report accusing colleagues of not paying enough attention to the issues being raised by animal welfare advocates.[2] More recently, a disapproving reviewer quoted this "fairly typical outburst" from a scientist who apparently still thinks the subject *ought* to be ignored: "There is a debate now as to what is the right of a mouse. Why are we wasting time in Washington with taking seriously this business?... This is complete and absolute craziness." The scientist who reportedly said this was none other than James Watson, codiscoverer of the DNA helix, in a speech given in 1985.[3] It remains an open question whether the research community is taking "this business" seriously *enough*, but one thing is clear: The number of articles in the scientific literature devoted to various aspects of the controversy *has* risen, and it has done so dramatically. In 1985, the latest year for which we have data, the number of such items found in five journals was almost 10 times larger than the number found in the same journals in 1977, the year of the FAS report. (See Figure 4.1. Our criteria for selecting relevant items and procedures for locating them are described in Chapter 3.)

For the moment, we will confine our discussion to the five journals searched by hand; afterwards, the analysis will be expanded to include data generated by computer searches of the entire scientific literature.

The annual data for the five journals, selected to represent different fields of science, suggest that since about 1982 the interest level in animal welfare topics has soared to a point well above any previous high mark within the past 2 decades. The passage of the Laboratory Animal Welfare Act in 1966 set off a flurry of news reporting and commentary that lasted

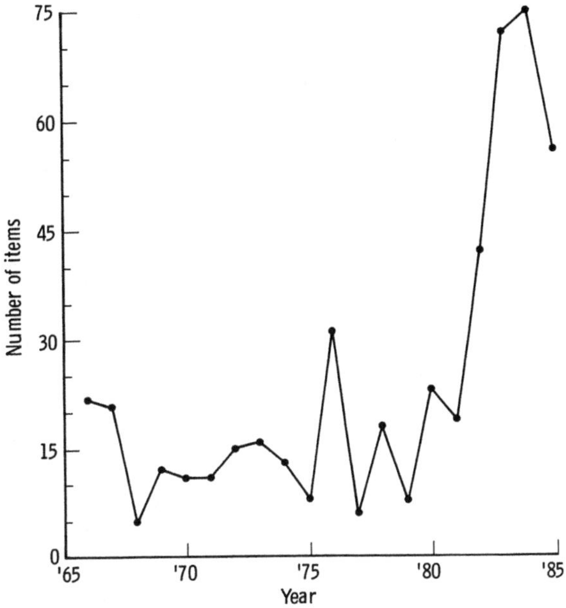

FIGURE 4.1. Number of research animal-welfare-related items published in selected journals, 1966–1985. This figure shows the combined totals of research animal-welfare-related items found each year from 1966 through 1985 in *American Psychologist*, *Journal of the American Veterinary Medical Association*, *New England Journal of Medicine*, *Physiologist*, and *Science*.

into the following year, then subsided abruptly. The level set in 1966 was not reached again until 1976—a year that saw the enactment of controversial amendments to the Animal Welfare Act (the word *laboratory* in the title had been dropped in 1970). Inciting more commentary than the 1976 amendments, moreover, was a series of demonstrations that same year by animal rights groups protesting experiments on cats at the American Museum of Natural History in New York.

Once again interest dropped off sharply, and rose thereafter only in fits and spurts until 1982, when it began a climb that set a new record in each of the next 3 years. These were years marked by a step-up in laboratory break-ins by animal rights activists, most notably an incident at the University of Pennsylvania in 1984. In that raid, activists identified with a group called the Animal Liberation Front stole videotapes of baboons undergoing head trauma experiments. The tapes were shown on national news broadcasts and prompted an NIH investigation of the laboratory that resulted in loss of funding for the research (see Chapter 2, fn 10). This was also a time of hectic legislative activity, as hearings were held on dozens of bills that eventually led to the passage of the Health

Research Extension Act and the Animal Welfare Act Amendment of 1985. The furor over these events has dwarfed the fuss in 1966 over the original Laboratory Animal Welfare Act. (See Appendix D: Chronology of Significant Events, 1966–1989.)

Conspicuously missing is any marked increase in items in the year that the 1970 amendment to the Animal Welfare Act (PL 91–579) was enacted. This cannot be entirely explained by the uncontroversial nature of the act. Its major provision was unobjectionable from the scientists' viewpoint: It merely extended coverage to animal exhibitors (eg, circuses and zoos)—hence the removal of the word *laboratory* from the title. Scientists had long been complaining that they were singled out unfairly by the 1966 act. However, the 1970 act also extended coverage to more research facilities and more species, and it added a requirement that "appropriate use" be made of pain-relieving drugs, as well as a requirement that facilities submit an annual report to the United States Department of Agriculture. (See Chapter 2 for further details.) Nevertheless, it was considered harmless enough for endorsement by the National Society for Medical Research.[4]

As originally proposed, the 1970 legislation had contained a provision that was (and still is) anathema to the research community: extension of the act's coverage to include regulation of the experimental procedures themselves. This did precipitate some cries of outrage, but not many found their way into the scientific literature.[5] Scientists were not particularly concerned, it seems, or very interested (except for veterinarians, as noted below). One reason for this apathy is that animal welfare activists had not yet turned the issue into the public cause célèbre that it became a few years later. Animal rights groups were not highly organized in 1970, not yet effective in orchestrating protests and laboratory raids, nor yet adept at writing their own bills as they began to do in the 1980s. Even without a massive public campaign, scientists were able to defeat the hated provision for interference with experimental procedures. The big difference between the relatively ho-hum response in 1970 and the outpouring in print when the 1976 and 1985 amendments came under consideration was fear—fear of what antivivisectionists might accomplish through their high-pressure tactics with Congress.

The total number of relevant items found in the five journals was down to 56 items in 1985, compared to 72 and 75 for the previous 2 years, respectively.[6] The number for any one year may not be particularly significant, though, since annual fluctuations are the rule for the entire period. The continuous rise from 1982 through 1984 is, in fact, the only instance of a trend either upwards or downwards that continued in the same direction for more than 2 years straight. For this reason, it is a little misleading to compare individual years, as we did, for instance, when we noted that the number of items found in 1985 is nearly 10 times greater than in 1977. That is true, but it ignores the fact that 1977 represents a low point sandwiched between two much higher years.

Let us consider, then, *average* numbers of items found over periods of 4 years. By taking the average number of items found annually during each 4-year period of time, we can smooth out the yearly fluctuations and see the general trend more clearly. The results are shown in Figure 4.2. Between 1966 and 1969, an average of 15 items were found per year in the five journals combined; for the next 4-year period, an average of 13.25 items were found per year; and so on, to the period 1982–1985, when an average of 61.25 items were found per year.

Looking at 4-year periods, one can see that the number of items found from 1966 through 1981 was relatively constant, at a level about one-fourth that of the latest period for which we have data. We would have considered a doubling or tripling of the number of items to be enormous; that the number is *four* times greater, even using this conservative measure, suggests that something momentous is afoot.

The data considered thus far are taken from only five journals, but as noted earlier, the trend remains generally the same when we expand the scope to examine the complete data from all sources: from the five selected

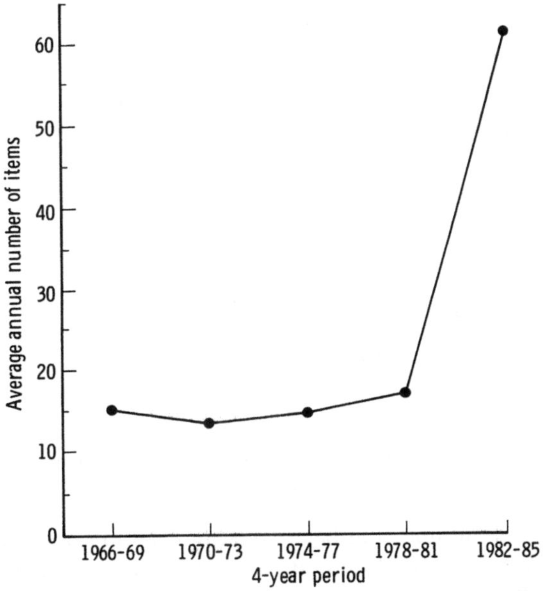

FIGURE 4.2. Average number of research animal-welfare-related items published in selected journals during four-year periods, 1966–1985. This figure is based on the same data as Figure 4.1, but the annual totals of relevant items found in the five selected journals are averaged over four-year periods in order to show the trend more clearly.

journals, plus all other items found through a computer search, with the addition of citations found in footnotes, bibliographies and other sources. The complete data, giving annual averages for 4-year periods, are presented in Figure 4.3.

Comparing Figure 4.2 with Figure 4.3 prompts two observations: First, it is still the case that the number of items found in the latest period (1982–1985) is roughly four times greater than the number of items found in the period 1966–1969. The difference, though—and this is the second observation—is that the 16-year period between 1966 and 1981 no longer appears as flat as it did in Figure 4.2. In fact, the complete data show a significant rise in the period 1978–1981, when the average of 34 items per year exceeded those found for the 1966–1969 period by over 50%. This indicates a more gradual rise in levels of interest in the animal welfare controversy than the pattern found in the five journals alone. The complete data suggest that the explosion seen in the 1982–1985 period was in fact preceded by a great deal of preliminary activity.

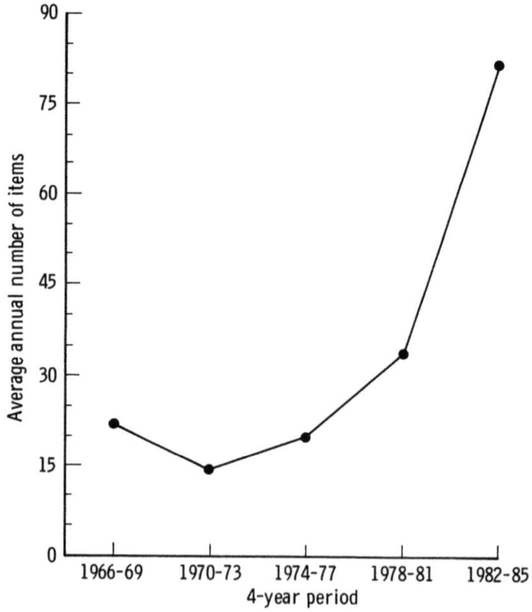

FIGURE 4.3. Average annual number of research animal-welfare-related items published in the scientific literature in the United States during four-year periods, 1966–1985. This figure shows the growth in the number of relevant items published in the scientific literature as a whole. Each point on the graph represents the average annual total for the four-year period indicated. Included in these data are all items from the five manually searched journals, *plus* all items found in a computer search of the literature (including articles published in books and full-length books).

The primary explanation for this discrepancy is that from 1978 through 1981 a wider range of journals took part in the debate. We found items in 27 different journals from 1978 through 1981, compared to 16 during the previous period, and only 8 in the period before that. From 1982 through 1985, 45 different journals published relevant items; this is more than double the number of journals for the period 1966–1969, and more than five times the number of journals for the period 1970–1973 (Figure 4.4). By this measure, the current high level of interest in the animal research controversy came about gradually, rising fairly steadily from a low point during 1970–1973.

Another reason why the increased activity from 1978 through 1981 was not evident in the data from the five selected journals is that some of that increase was due to books and articles in books. A rise in these kinds of nonjournal items occurred a little earlier than did the big increase in the journals (Table 4.1). The complete data include books and articles in books and thus reflect this earlier increase, whereas the data from the five journals do not.

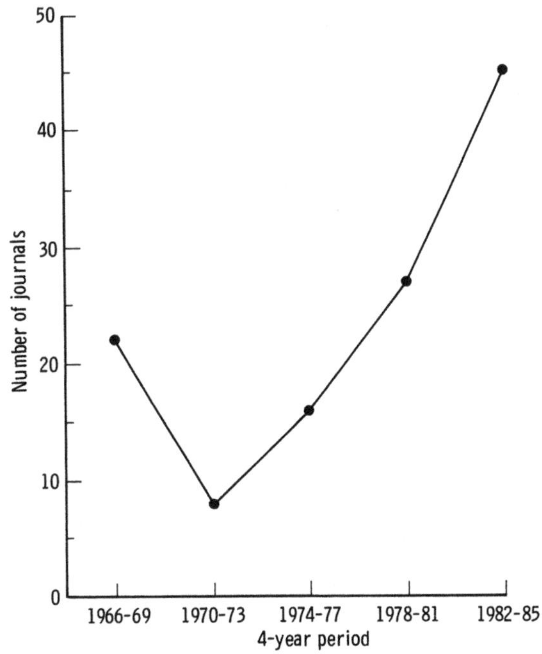

FIGURE 4.4. Number of scientific journals published in the United States in which research animal-welfare-related items were found, 1966–1985. This figure shows the total number of different journals in which research animal-welfare-related items were found during each four-year period from 1966 through 1985. (Articles in books and full-length books are not included in these data.)

TABLE 4.1. Publication of full-length books and articles in books on the welfare of research animals, 1966–1985.

	Articles in books	Full-length books
1966–1969	0	1
1970–1973	2	0
1974–1977	3	3
1978–1981	9	3
1982–1985	4	6

TABLE 4.2. *For five selected journals*, the number of journals publishing relevant items and the average number of relevant items per journal: Comparison between the periods 1966–1969 and 1982–1985. (The values in the table are cumulative totals for the 4-year periods indicated at the head of each column.)

	1966–1969	1982–1985	Percent change
Number of journals publishing relevant items	3	5	+67%
Number of relevant items per journal	20	49	+145%

TABLE 4.3. *For the scientific literature as a whole*, the number of journals publishing relevant items and the average number of relevant items per journal: Comparison between 1966–1969 and 1982–1985. (The values in the table are cumulative totals for the 4-year periods indicated at the head of each column.)

	1966–1969	1982–1985	Percent change
Number of journals publishing relevant items	22	45	+105%
Number of relevant items per journal	4	7	+75%

The foregoing analysis of the number of journals that published relevant items each year has a corollary: the number of *items* published *per journal* each year. Almost by definition, the latter is what accounts for most of the huge increase found in the five manually searched journals (ie, the number of different journals could not exceed five, although that number could, and did, drop below 5 during periods when one or more of the selected journals published no relevant items). (Table 4.2).

For the scientific literature as a whole, however, the increase in the number of different journals that published items on laboratory animal welfare outweighs the increase in the number of items per journal (Table 4.3).

The number of different journals that published relevant items could be considered an indication of the *breadth* of interest in research animal welfare. Likewise, one could take the number of items per journal as an indication of *intensity* of interest in the subject. A few journals publishing a great many articles would then suggest intense interest in the issue within a rather narrow segment of the scientific community. Conversely, the same number of items scattered among a large number of journals would indicate a less intense level of interest distributed more broadly throughout the community.

Using this terminology, we conclude that among the five manually searched journals, interest in the subject by the mid–1980s had more than doubled in intensity since the late 1960s. It had also broadened, from an initial three journals to involve, eventually, all five (Table 4.2). To get a real idea of breadth, however, one must look at the complete data. When the artificial upper limit imposed by the sample of five is removed, breadth of interest shows an increase of well over 100% (Table 4.3). For the scientific literature as a whole, more growth has occurred in breadth than in intensity: The same journals are increasing their coverage of the subject, but this increase in intensity is outweighed by the increase in breadth. The subject is involving a wider and more diverse audience than ever before.

The period from 1970 through 1973 illustrates the fact that breadth and intensity do not always fluctuate in the same direction. During this period, the overall number of published items fell slightly (Figure 4.3). This was because the number of different journals publishing relevant items dropped, but the extent of that decline was masked by a rise in the number of items per journal. In other words, interest in the subject narrowed down to a very few diehards, who stepped up their writing to a level of intensity that matches the mid–1980s. The hot spot was the *Journal of the American Veterinary Medical Association*, which published 47 of the 57 items found from 1970 through 1973. Veterinarians were (and still are) intensely interested in laboratory animal welfare issues, especially legislation, but at that time virtually no one else was.

Up to this point, no distinction has been made between items of greater or lesser significance. In the figures above, a one-paragraph news brief counts the same as a feature article, or even a full-length book. The sense of what we are trying to measure might have been better conveyed by counting a collected volume of, say, 20 papers as 20 items. The problem with that method is that such a book would weigh in 20 times heavier than a book written by only one author. (Articles in books that were not wholly concerned with relevant issues were counted separately, as explained in Chapter 3.) Rather than devise a complicated weighting system to try to account for differences in length or importance among items, we opted for simplicity. In one sense, then, the figures understate our thesis about the growth of the debate over the past few years in the scientific literature. This is true because most of the full-length books have been published quite recently. Of the 13 full-length books we found on the subject, 6 have been published since 1982 (Table 4.1). The recent appearance of these more substantial contributions to the literature further supports our thesis that the debate at present has taken on important new dimensions.

But what about the nonbook items? It would make a difference if, for instance, the increase in recent years is due entirely to news briefs. It could be that the publication of feature articles—items most likely to contain in-depth discussions of the issues—actually became scarcer while news flashes proliferated. That eventuality would lead to the suggestion that the debate became shallower, even as it increased in breadth and intensity.

There is, in fact, some indication that things are heading in that direction. To sort out separate trends, we divided items into five categories: feature articles, news briefs, letters, books, and other (which includes such miscellany as committee reports, editorials, book reviews, and guidelines). The average annual number in each category for the 4-year periods from 1966 through 1985 is presented in Figure 4.5.

It is now obvious that the overall increase in relevant items is *not* entirely due to news briefs. All categories show a large increase in the period 1982–1985. However, news briefs—those little notices about relevant legislation found in many professional journals—proliferated much more rapidly of late than did feature articles. News briefs accounted for about one-fourth of all items during 1982–1985. They had accounted for only one-sixth during the previous period, and they moved in ranking from last place (not counting books) to first place. This suggests that legislative concerns have overtaken other kinds of interests in research animal-welfare issues.

Another way of making the point is to note that during 1978–1981, feature articles comprised almost half of all items (61 out of a total of 136 items), compared to only a quarter in 1982–1985 (80 out of 327 items). This means that even though many more relevant items were found in the 1982–1985 period, a smaller proportion of them contained sustained discussions of the issues.

One other interesting revelation is that readers—the people who write letters to the editor—have become much more interested in the controversy of late. Through 1977, most of the published items were of the type generated by editors. Only after 1978 did readers begin to get into the act, writing letters on the subject to scientific journals at a quickly accelerating pace. Letters provide the most dramatic illustration of increased activity: From a low point of only three letters found during the 4 years from 1970 to 1973, their incidence rose to 80 during 1982–1985. (Of course, we do not know how many letters may have been written but not published, so this may be more a measure of editors' willingness to publish letters on the subject than of readers' readiness to write them.)

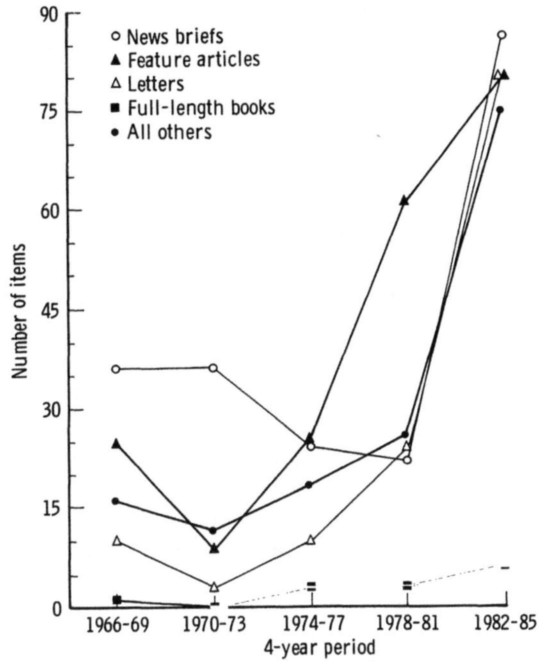

FIGURE 4.5. Comparison of the numbers of research animal-welfare-related items in different categories published during four-year periods, 1966–1985. The values given in this figure represent totals for each four-year period, not annual averages. The categories that are compared are based on the format rather than the content of the published items.

The final quantitative question we will consider arises from the fact that a disproportionate percentage of all items came from one source, the *Journal of the American Veterinary Medical Association* (*JAVMA*). Of the total of 690 items found in all sources (including articles in books and full length books), 298, or about 43%, are from this one journal. If journal items alone are considered, 45% are accounted for by *JAVMA*. It could be that all our conclusions really apply only to *JAVMA*, and that the sheer weight of data from this one journal skewed the composite findings. Perhaps what is true of veterinarians is far removed from what is happening in the rest of the animal research world.

To examine this possibility, we first separated *JAVMA* items from all others and redrew the line graph depicted in Figure 4.3. The results (not shown) were similar enough to the original figure to dispel doubts on that score. However, the *JAVMA* data did show a sharper rate of increase in 1982–1985 over the previous period than did the rest of the data.

Taking the question one step further, we investigated the role played by all the veterinary journals in which we had found items, compared to journals in other scientific fields. The 79 journals in which items had been found were divided into four categories: biomedical (54 journals), veterinary and animal science (11), psychology and behavioral science (6), and general science (8). (For a complete listing of all journals by category and the number of items found in each journal, see Appendix E.) The rate of increase for each category of journal is plotted individually in Figure 4.6.

Again, our general conclusions are confirmed. In all fields, the number of relevant items found increased greatly during the 20-year span of time covered in our survey. Comparing the volume of the mid-1980s to the period of ferment over the first federal legislation in 1966, we see the following (rounding off to the nearest whole number): veterinary and animal science journals—3 times more items found during 1982–1985 than during 1966–1969; biomedical journals—5 times more; general science journals—4 times more; psychology and behavioral science journals—a startling 11 times more (because this category started out with only 1 item, an addition of just 10 items produces a deceptively large multiplier). Throughout, veterinarians have shown consistently higher

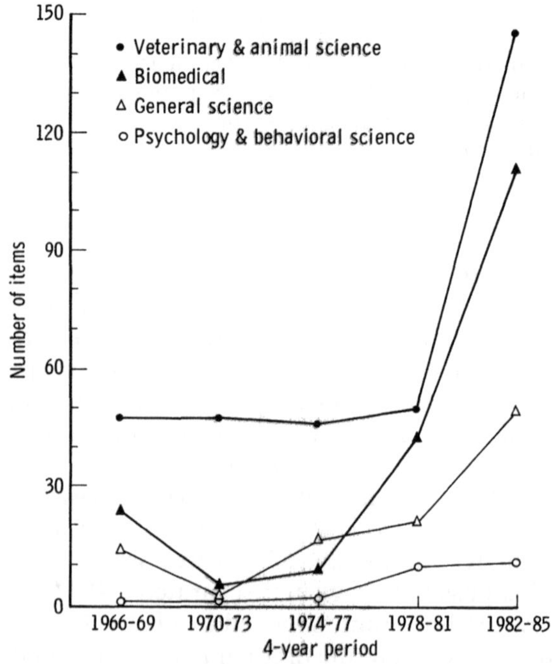

FIGURE 4.6. Comparison of the numbers of research animal-welfare-related items published in different types of journals during four-year periods, 1966–1985. The values given in this figure represent totals for each four-year period, not annual averages. (Articles in books and full-length books are not included in these data.)

levels of interest in the topic (at least, in terms of *intensity*) than has anyone else.

The following chapter will deal with the content of the articles, letters, editorials, and other expressions of opinion that lie behind the numbers we have been scrutinizing. Before moving on to qualitative matters, however, let us summarize what we take our findings to mean thus far. First, we have documented and quantified what everybody already knew about the current high level of interest in research animal welfare issues. Toward the end of the 20-year period of the study, scientists in this country began reading and writing about the animal research controversy in their professional literature much more often than ever before. This conclusion is upheld, whether one looks at data from five selected journals or from a computerized search of the entire literature. It is also the case within each of four scientific fields examined separately.

Hundreds of articles and other kinds of items on the topic appeared in the scientific literature from 1982 through 1985, but the explosion was not as sudden overall as it appeared in the five selected journals alone. During the period 1978–1981, the debate expanded in breadth, spreading beyond the pages of the few journals that had always been interested in laboratory animals. The numbers of feature articles rose most rapidly during this period; and among biomedical and psychology journals, this period was the time of greatest growth.

Further, we considered some aspects of this heightened interest that are not immediately apparent to the casual observer. These are different components of interest in research animal welfare, which we labeled breadth, intensity, and depth of interest. All three are now demonstrably greater than during most earlier periods, but the increase in breadth is most striking. The present level of intensity, as measured by the average number of items found per journal, was matched once before, in the period 1970–1973. Depth, as measured by the proportion of feature articles in the total of all items, is falling off again. (On the other hand, full-length books, which also indicate depth, are on the rise.) Breadth, as measured by the number of different journals in which items were found, soared in the period 1982–1985. The steady expansion of the debate to more and more journals has imparted a unique character to the controversy today. A larger segment of the scientific community is taking part in the discussion. In addition, a great many scientists have just recently taken to writing letters on the subject to their professional journals.

Linking the quantitative data with important events in the public arena shows that the level of the scientific community's interest in animal welfare issues is very sensitive to political developments. Legislation, or the threat of it, caused the number of relevant published items to rise in 1966, 1976, and in the mid–1980s. These years marked the dates of the Laboratory Animal Welfare Act of 1966 and the amendments of 1976 and

1985. Public demonstrations and raids coincided—not at all coincidentally—with legislative activity during these peak years.

It was suggested that the ascendance of news briefs to the most numerous category during the period 1982–1985 indicates that scientists' interest in animal welfare has become more narrowly focused on legislative concerns. However, this cannot be judged accurately from the data presented in this chapter. Therefore, one of the questions to which we will turn in Chapter 5 is, Does the scientific literature indicate a willingness to discuss the moral and ethical issues that have been raised by critics, or is this effluence merely defensive? More generally, what themes emerge in the literature, and how have they changed over the past 20 years?

Notes

(The notation *[AB]* after a reference indicates that it is included in the Annotated Bibliography following Chapter 7.)

[1] Cowley G et al (1988) Of pain and progress. *Newsweek 112*,26(Dec26):50–57.

[2] Federation of American Scientists (1977) Animal rights. *Pub Interest Rep 30*,8:1–8.

[3] Cartmill M (1986) Animal rights and wrongs. *Nat Hist 95*,7:66–69.

[4] Visscher MB (1971) The Animal Welfare Act of 1970 (P.L. 91–579). *Science 172*:916–917 [AB].

[5] Public testimony stating the opposition of the American Veterinary Medical Association to this provision of the bill can be found in Decker WM (1970) Laboratory animal care legislation. *J Am Vet Med Assoc 157*,3:284–285 [AB].

[6] Although we have no data beyond 1985, there are no indications that interest in the subject has fallen off since that year. A resurgence of interest among scientists has been virtually guaranteed by the aggressive tactics of animal protection groups, which have been increasingly successful in halting animal research projects by forcing funding agencies and institutions to withdraw support. Two events in 1988, neither of which involved break-ins or other illegal activities on the part of animal rights activists, illustrate this trend. The National Institutes of Health (NIH) responded to criticism from the Physicians Committee for Responsible Medicine, as well as from animal protectionists, by stopping funding for a $1 million project to study head trauma in cats at the University of Cincinnati College of Medicine. And in New York, a researcher at Cornell Medical College returned a $600,000 grant to the National Institute on Drug Abuse (NIDA) after protesters criticized her experiments studying barbiturate addiction and withdrawal in cats. The official story is that the researcher returned the grant voluntarily, but news reports indicate that Cornell pressured her into it. (See Gribetz S [1989] Newsfronts. *Lab Anim* (NY) *18*,1:8–9 and Lyall S [1988] Pressed on animal rights, researcher gives up grant. *New York Times* (Nov 22):B1,B5.) In addition, since 1985 there have been numerous instances of thefts of laboratory animals and vandalism conducted

against animal research laboratories, for which groups such as the Band of Mercy or the Animal Liberation Front have claimed responsibility (see Appendix D). These developments may not contribute to scientists' willingness to consider the ethical complexities of the matter, but they do tend to focus the attention of the scientific community on the controversy.

CHAPTER 5

Qualitative Analysis: Viewpoints and Opinions

Anyone glancing through the scientific literature on the laboratory animal controversy would quickly notice how very much of it is devoted to protests against the regulation of research. One could not read far before encountering certain ubiquitous themes used in support of those protests: the importance of animals to research, the contributions of animal experimentation to human health, the view that animals are already humanely treated, and a perception that critics of animal research are uninformed and emotional. These themes were encountered again and again over the 20-year span of the survey.

No doubt about it, legislative or regulatory concerns were foremost in the minds of most of the authors included in this study. That is not the whole story, though. Other topics that have gained attention within the last decade include moral or ethical issues of animal research; the problems of identifying and alleviating animal pain, other forms of distress, and suffering; and prospects for the development of alternative methods. Together, these four topics account for the content of the vast majority of items.

In order to quantify these general impressions and to analyze changes in emphasis over the years, we coded each item by topic. In addition, we noted the overall tone of the author: either defensive, as the themes mentioned above would suggest, or critical of animal experimentation. Finally, we identified dozens of themes expressing positions on specific issues. By tabulating the frequency of each theme's appearance, we were able to chart shifting opinions and attitudes from 1966 to 1986.

In the following pages each of these matters—topics, tone, and themes—will be considered in turn.

Topics

Insofar as the scientific literature has dealt with animal welfare issues at all, the overriding catalyst for doing so has always been the threat or the reality of restrictions. *Science*, the primary news magazine for the scientific community, has devoted countless news briefs and dozens of feature articles to animal research developments that continue to unfold in Congress and in the government's regulatory agencies.[1] The same is true, on a smaller scale, of *BioScience*, the official publication of the American Institute of Biological Sciences.[2] Moreover, some of the specialty journals now have regular columns to advise readers of the progress of animal welfare bills making their way through Congress. The *Journal of the American Veterinary Medical Association* (*JAVMA*), for instance, has a Washington news department that has included a laboratory-animal welfare section in almost every issue for 20 years. In 1981 the *Physiologist* inaugurated a "Public Affairs" column "to keep the membership informed of legislative and other issues affecting animal experimentation."[3] The column has served as a forum to publicize testimony given by scientists at Congressional hearings as well as a source of information about legislation on both national and local levels. Other journals publish similar departments intermittently; an example is *Cancer Investigation*'s "Public Issues," which has on occasion been devoted to research animal legislation.[4]

Over half (60%) of all 717 items included in this study were concerned primarily with matters of regulation. From 1982 through 1985, three times as many items on this topic were found (200 items) than in the previous 4-year period (66 items). It is characteristic of the contemporary phase of the debate, however, that the subject is no longer the *only* one under discussion. In earlier years, items focusing on regulatory concerns comprised 80% or 90% of the total; this proportion has dropped sharply since the mid–1970s (to 58% for the period 1982–1985), even though the number has risen (See Table 5.1).

The surprise lies in our discovery that the scientific literature has also opened up, to a considerable degree, to discussions of the moral arguments raised by animal welfare proponents. Scientists cannot ignore legislation or other regulations that restrict their work, but they can and did for a long time ignore the ethical questions raised by critics. Table 5.1 shows that prior to 1974 only three items (out of 139 found for 1966–1973) were concerned primarily with ethical or moral issues. In contrast, 73 discussions of ethical issues appeared during 1982–1985, comprising 21% of the 344 relevant items found for that period.

Attitudes typical of the early years were expressed more baldly than usual in a paper presented in 1969 at a symposium sponsored by the National Academy of Sciences' International Committee on Laboratory Animals. The author made the point that animals should be substituted for humans in experiments wherever possible. The advantage, he wrote, is that "on animals all we have to do is to take safeguards about pain, and then we can do what we wish on the animal, but on humans this is not so." Furthermore, he added, the Helsinki Declaration on Human Experimentation states that "'every clinical research project should be preceded by a careful assessment of inherent risks in comparison with

TABLE 5.1. Research animal topics in the scientific literature in the United States, 1966–1985: Changing concerns.[a]

Articles or books concerned primarily with	1966–1969	1970–1973	1974–1977	1978–1981	1982–1985	Total for 1966–1985
1. Legislative or regulatory issues	83% (69)	91% (51)	54% (44)	43% (66)	58% (200)	60% (430)
2. Ethical or moral issues	4% (3)	0	17% (14)	34% (52)	21% (73)	20% (142)
3. Pain, other forms of distress, and suffering	5% (4)	4% (2)	19% (15)	5% (7)	5% (16)	6% (44)
4. Alternatives to the use of animals in research	0	0	9% (7)	16% (24)	6% (22)	7% (53)
Total number of items upon which the percentages are based	83	56	81	153	344	717

[a]Percentages for each column do not total 100% because some articles did not deal with any of these topics and a few were equally concerned with more than one.

the foreseeable benefits to the subject and others.' There are also special precautions about ... what is going to happen after the experiment. None of this applies to animals."[5]

There was general agreement on these points, as the author remarked in his paper. Few people at that time thought of applying ethical concepts to animal experimentation. When they began doing so in the late 1970s, many scientists were taken by surprise; so much so that a pharmacologist thought (in 1983) that ethics must be a "relatively new discipline"—a notion that would certainly have dismayed Aristotle. In his fervent support for the continued use of animals in research, combined with a cautious acceptance of ethical considerations, the pharmacologist expressed a common reaction to the new ideas:

> [T]he abrupt discontinuation of animal research would have a terrible impact on the life sciences, but there is no reason why workers should not try to be humane in their handling of laboratory animals. The relatively new discipline of ethics has contributed importantly to an awareness of the proper way to conduct human research. There is every reason to believe that the discipline of ethics can contribute equally to an understanding of acceptable ways to do animal research.[6]

One of the three items that did contain a discussion of ethics prior to 1974 was a heated response to an antivivisectionist tract claiming that animals have rights; the author countered that nothing should take priority over human welfare.[7] The other two items—one an article, the other a book of essays—foreshadowed the critical concern with ethics that would emerge shortly in the scientific literature.[8] By 1977 that concern was so evident that a writer referring to animal experimentation could state without contradiction, "It is now *de rigueur* to include some observations on legal or ethical aspects in comprehensive discussions of biomedical matters. A generation ago, this feature would hardly have been considered."[9] His words were to become even more true in the years to follow.

Nevertheless, ethical and moral concerns have not been institutionalized in the way that legislative concerns have been, through regular columns devoted to the subject in scientific journals. A letter published in the *JAVMA* in 1983 suggested such an idea,[10] and the proposal was promptly seconded,[11] but nothing came of it. Nor does it seem likely to happen in the future. Morality remains a subject that most scientists are wary of, and in fact the proportion of items in this category fell in 1982–1985 from its peak during the previous 4-year period.

A third general concern that emerged was the problem of identifying and alleviating animal pain and other kinds of suffering or distress. Many articles included in the first two categories mentioned pain and suffering, but only in abstract terms. Authors included in this third category took a concrete approach to the subject; they were interested in the practical problems of relieving distress in animals, without dwelling on the philosophical or regulatory issues involved. Few authors—mostly veterinarians—have tackled the subject, but the number has grown.[12]

During the period 1974–1977, the pain-and-suffering category rose to encompass 19% of the 81 items found (compared to 6% overall). The explanation for the sudden surge of interest reveals, once again, how strong a motivator is the long arm of the law. The 1970 amendment to the Animal Welfare Act had added a requirement for "adequate veterinary care, including the appropriate use of anesthetic, analgesic, and tranquilizing drugs, when such use would be proper in the opinion of the attending veterinarian."[13] The reaction was slow in coming, but in January 1975 a directive from the United States Department of Agriculture put renewed emphasis on veterinarians' responsibilities in relieving pain and distress.[14] That summer the American Veterinary Medical Association organized a symposium entitled "Adequate Veterinary Care" at its annual meeting. Abstracts of some of the papers presented at the symposium were published in the *JAVMA* a few months later, and the papers were published in full the following year.[15] This one symposium accounted for nearly half of the 15 items included in this category for 1974–1977.

It would appear from the figures in Table 5.1 that interest in the practical problems of pain alleviation dropped after 1977; during the following 8 years, the proportion in this category returned to the 5% level, where it had been earlier. However, several of these later works are much more substantial than the earlier articles. A full-length book published in 1980 focuses on the problem of how to recognize animal suffering.[16] One of the most comprehensive treatments to date is a volume published as a result of a symposium in 1982 sponsored by the Federation of American Societies for Experimental Biology (FASEB) and the American Veterinary Medical Association (AVMA). The symposium, entitled "Animal Pain: Perception and Alleviation," was "the first symposium known to us that has specifically addressed pain in animals," according to the editors of the proceedings.[17] As noted in Chapter 2, the 1985 amendment to the Animal Welfare Act and the new PHS Policy have now given a legislative mandate to institutional review committees to consider animal pain and distress in their evaluations of research practices. A substantial report addressing this issue was published by the American Veterinary Medical Association in 1987, the proceedings from a colloquium entitled "Recognition and Alleviation of Animal Pain and Distress."[18]

The fourth topic, alternative methods, was missing altogether from the first decade of the literature surveyed in this study. The concept, but not the term, was mentioned briefly in a 1966 paper cited previously as one of the first to use an ethical framework for evaluating animal research. The same author dismissed as "a vain hope" the idea that "lower forms of life" could replace laboratory animals.[19] The next reference to alternatives—again without using the term—came in a 1971 editorial in the *Journal of the American Medical Association.* This time the range of alternative methods was broadened to include tissue culture techniques, mathematical models, and computer simulation, but the author was no more optimistic than the earlier author had been. The Fund for the Replacement of Animals in Medical Experiments (FRAME), which had been founded in England in 1969, "might be better called 'FRAUDS' (Fund for Replacement of Animals Used in the Discovery of Science)," wrote the anonymous editorialist. "Just how these methods might substitute for animal experimentation in neurophysiology, for example, is difficult to comprehend."[20]

It was not until about 1976 that the term *alternative methods* entered the vocabulary of the debate on a large scale and received a sympathetic hearing. That year an entire chapter on alternatives was included in a book criticizing painful animal experiments,[21] and the work was revised 4 years later to make alternatives its main focus.[22] In 1976, also, the National Academy of Sciences (NAS) sponsored a symposium to consider alternatives. Although many observers thought that the emphasis was more on the necessity for continued use of animals in research than on the development of alternatives,[23] the symposium did open up the topic for more general discussion. At least one paper presented at the NAS symposium reminded the audience of the "three Rs": *r*eplacement, *r*eduction, and *r*efinement in animal experimentation.[24] First advocated in a 1959 British book on experimental technique written by W.M.S. Russell and R.L. Burch, the three Rs have played an important role in the current understanding of the meaning of *alternatives*. Recognizing this, a writer in 1981 took the opportunity to review the Russell and Burch volume more than 20 years after its publication by including the review with one of a contemporary book on the subject of alternatives.[25] An alternative, an author of the latter book wrote elsewhere, is defined as "any technique that could *replace* the use of animals altogether, *reduce* the number of animals required, or, through *refinement* of the techniques used, reduce the amount of distress that the animals suffer."[26]

In the early 1980s, alternative methods constituted a particularly hot topic. As usual, the scientific literature responded to public events: At least three bills designed to promote the development of alternatives were being debated in Congress at the time.[27] A coalition of animal welfare groups had launched a campaign in 1979 to find alternatives for the Draize eye irritancy test used in the production of cosmetics and shampoos,[28] and their activism paid off in the donation of several large grants for programs to develop alternative methods of testing[29] (see Appendix D: Chronology of Significant Events). In the midst of all this, the National Institutes of Health held another conference on alternatives.[30] Most significantly, in 1985 the concept of alternatives was translated into federal law for the first time (for details, see Chapter 2).

Relatively speaking, attention given to alternative methods in the scientific literature has lessened since 1981. Table 5.1 shows that the number of articles on the subject remained about the same during 1982–1985 as it had been during the previous 4 years, but that number constituted a smaller proportion of all items.

Tone

It is important to know what the concerns of scientists are, but the information is ambiguous without also knowing what stance scientists take toward those issues. Shortly we will turn to our findings on specific positions regarding eight separate topics, but first let us consider stance in terms of the overall tone imparted by a piece of writing.

About half of all items included in this study (362 out of the total of 717) clearly "take sides"; that is, they defend animal research, or they attack some aspect of it. Most of the critical papers include an admission that animal research is justified under some circumstances, so papers classified as "critical" are not necessarily antivivisectionist, nor are they all wholesale denunciations of animal research. In fact, our criterion for a "critical" item was quite lenient, requiring only that the author suggest a need for change. Examples include authors who find fault with the extent of legislative protection currently afforded laboratory animals,[31] enforcement of those protections,[32] the conditions under which laboratory animals are kept,[33] the lack of attention paid to ethical questions and to the development of alternative methods of research,[34] specific practices such as multiple surgery for training purposes,[35] and so on.

In the 1960s and early 1970s the number of critical articles published was small compared to the number of articles justifying animal experimentation (see table 5.2). For the next 8 years, critical articles outnumbered the purely defensive ones. Most recently, a balance has been struck: During 1982–1985, each camp occupied about 28% of the territory (with the remaining 44% not explicitly on either side).

Dating from the mid–1970s, examples of the new critical stance abound. In 1976 a psychiatrist, Dallas Pratt, wrote a book-length exposé, a harsh indictment charging the research community with inflicting unjustified pain and distress on animals; a new edition was issued in 1980.[36] The Pratt books became required reading in animal welfare circles, but it is unlikely that a large segment of the scientific community has read them. Much more influential among researchers was a 1977 report by the Federation of American Scientists accusing the National Society for Medical Research—an organization founded in 1946 expressly to defend animal experimentation—of being "so locked in combat with antivivisectionists that it ignores the possibility that the treatment of laboratory animals might be improved without impairing laboratory research."[37] This article has been credited with at least one palpable result: the founding in 1978 of the Scientists Center for Animal Welfare (SCAW), a Washington, DC-based group of scientists that works to improve conditions for laboratory animals.[38]

The same year that SCAW was organized, a letter in *Science* noted that "a surprising number of scientists ... are increasingly questioning the scientific and moral basis upon which even the most enlightened animal experiments rest." The author added his opinion that "the broader sci-

TABLE 5.2. Tone of research-animal items found in the scientific literature in the United States, 1966–1985: Justifications and criticisms.[a]

Overall tone of items	1966–1969	1970–1973	1974–1977	1978–1981	1982–1985	Total for 1966–1985
Justifications	31%	16%	20%	24%	28%	26%
	(26)	(9)	(16)	(37)	(95)	(183)
Criticisms	10%	7%	23%	33%	28%	25%
	(8)	(4)	(19)	(51)	(97)	(179)
Neutral	59%	77%	57%	42%	44%	50%
	(49)	(43)	(46)	(65)	(152)	(355)
Totals	100%	100%	100%	99%	100%	101%
	(83)	(56)	(81)	(153)	(344)	(717)

[a]Percentages in each column do not always total exactly 100% because of rounding errors.

entific community has not paid enough attention to such issues and questions."[39] A textbook for the experimental study of animal behavior published that year took the unusual step of including a chapter on ethics and even criticized the morality of some of the experiments cited elsewhere in the book.[40] The following year a neurologist who works with fish rebuked researchers for not taking seriously the pain they inflict on cold-blooded creatures.[41]

Since the beginning of the 1980s, critical voices in the scientific literature have been easier to find. The voices are not always those of scientists; occasionally they belong to outsiders from the animal welfare camp. Their writing is not in the tradition of mea culpa, but of *j'accuse*. Nevertheless, whenever a scientific journal opens its pages, or a scientific conference its podium, to animal welfare advocates, that in itself could be considered an act of self-criticism. There are many recent instances of this: philosopher Bernard Rollin's warning, published in a psychology journal, that "much work needs to be done to elevate the status of research animals,"[42] and his scolding delivered to veterinarians for ignoring animal welfare issues in their ethical principles;[43] an essay by another philosopher, Arthur Caplan, addressed to a scientific workshop, that discussed the moral value of animals and concluded with the injunction that "the burden of proof rests upon the experimenter to justify the use of animals in experimental contexts";[44] an interview with Henry Spira, organizer of the Coalition to Stop Draize Rabbit Blinding Tests, in *Lab Animal*;[45] Christine Stevens' (founder and president of the Animal Welfare Institute) comparison of the lot of laboratory animals with that of human prisoners in the Soviet Gulag, in speaking to a symposium organized by the National Academy of Sciences;[46] letters published in *Science* from John McArdle (Humane Society of the United States), urging the biomedical research community to accept that "change is needed and desirable"[47] and from the late Eleanor Seiling (United Action for Animals) defending a bill written by her organization to promote the use of alternatives to live animals in research.[48]

The scientific literature contains dozens of critical articles and letters written during the past decade by two well-known animal advocates who also speak as scientists. Andrew N. Rowan, a biochemist and an articulate spokesman for alternative methods and for ethical considerations in animal research, charged in 1981 that "in much of biomedical research, animals' suffering and their deaths are unnecessary."[49] Rowan's recent book on the subject is the most authoritative and credible of the critical evaluations of animal research.[50] Michael W. Fox,[51] a veterinarian, is another persistent critic who writes eloquently of the need for empathy and for more consideration for animals' psychological and emotional requirements.[52] Fox recently challenged "all members of the [veterinary] profession to take a more radical and affirmative stance in support of animals' welfare and rights."[53] He has also accused experimental psychologists of allowing "needless repetition and unnecessary (unjustifiable) suffering" in their use of animals.[54]

Many other scientists have joined the doubters. Some have singled out their own disciplines for censure, as a psychologist did in 1980 when he accused colleagues of "a serious omission" in paying so little attention to ethical questions.[55] Recent letters in the *Journal of Oral Surgery* and in *The New England Journal of Medicine* objected to the lack of ethical justification for procedures described in research reports.[56] An editorial in the *Journal of the Experimental Analysis of Behavior* lamented, "we never seem to question our treatment of laboratory animals and the brutalizing effect of some of our practices on ourselves and our students...."[57] And a veterinarian accused the American Veterinary Medical Association's (AVMA) Animal Welfare Committee of a whitewash in its 1983 annual report:

> The Animal Welfare Committee was established because the AVMA took heed that something was brewing in the wind regarding animal rights, involving pain, suffering, and wrongdoing to animals. And what does it find? That essentially almost nothing is wrong.[58]

There is a paradox here. We have cited these critical observations to make the point that the scientific community has begun to question its treatment of laboratory animals—and yet the very authors we cite argue that the scientific community is *not* facing these questions. Merely opening the gates to self-criticism, obviously, has not done away with the conditions that inspired that criticism. In fact, many scientists (like the AVMA committee in 1983) see no problem that needs correcting.

Of those who assume a defensive posture, many approach the debate with the single-minded purpose of demonstrating the benefits of animal research. Although this genre rarely addresses the concerns of would-be reformers, it has produced some excellent surveys of the wide variety of ways in which animals have proved valuable to medical progress. In response to the threat of the 1966 Laboratory Animal Welfare Act, for instance, the American Public Health Association and the Conference of Public Health Veterinarians organized a symposium for the express public relations purpose of delineating the contributions of laboratory animals to various health fields.[59] The papers provide a wealth of information on the early role of animal experiments in surgery,[60] nutrition,[61] drug development,[62] and veterinary medicine[63]—but not one makes even a passing reference to any ethical issues.

In 1974 Charles Best was asked by an editor at the *Physiologist* to write "a short essay on the importance of dogs in medical research," explicitly because of the large number of restrictive bills introduced that year in Congress. Best's article describes how he used dogs in his work with Frederick Banting that led to the discovery of insulin as a treatment for diabetes.[64] Several papers delivered at a symposium on animal research held the following year (but not published until 1977) also cataloged specific areas of medical progress made possible through animal experimentation,[65] as did the introduction to the published proceedings.[66]

This kind of justification is still popular in the 1980s. The value of animal models of human disease was the subject of a paper delivered at a 1980 seminar.[67] The theme was given a personal twist by a past president of the American Physiological Society in 1983 when he described, on numerous occasions, the animal experiments that had made possible his own coronary bypass operation.[68] In 1985, the latest year included in our study, the *Physiologist* launched a series of articles devoted to the "health benefits of animal research"; the first two contributions surveyed the experimental roles of rats and of primates, respectively.[69] Also in 1985 a major article in the *American Psychologist*, complete with dramatic before-and-after photographs of patients, described the value of the use of animals in behavioral research. The author's introductory abstract provides a summary of the enormous range of medical benefits discussed more fully in the text:

> Some of the significant contributions of animal research covered are those to welfare of animals; treatment of human urinary and fecal incontinence; psychotherapy and especially behavior therapy and behavioral medicine; rehabilitation of neuromuscular disorders; understanding and alleviating effects of stress and pain; discovery and testing of drugs for treatment of anxiety, psychosis, and Parkinson's disease; new knowledge about mechanisms of drug addiction, relapse, and damage to the fetus; treatment enabling extremely premature infants to gain 47% more weight and save $6,000 per child in hospital care; and understanding the mechanisms and probable future alleviation of some deficits of memory that occur with aging.[70]

Other types of defense deal more directly with the issues raised by reform-minded critics, who rarely dispute that animal research can be valuable. Some stress the high standards of care already afforded laboratory animals;[71] some argue the case for scientific freedom;[72] many (during the 1980s) belittle the possibilities for alternative methods;[73] and a few consist of attacks on the motives or intelligence of the critics.[74] One justificatory paper written in 1984 is particularly interesting, if atypical, because it is based on an empirical study. The authors conducted a survey of every animal experiment reported over the previous 5 years (1979–1983) in research journals published by the American Psychological Association, in order to refute accusations leveled against psychological research in an animal rights publication. The alleged abuses included administering intense, inescapable electric shocks and depriving animals of food and water until they died. Not one example of any of the abuses was found in 608 research reports. The authors concluded that if such abuses occurred at all,

> they certainly were infrequent, and it is extremely misleading to imply that they were typical of experimental psychology.... By misleading humane people, the

radical animal activists are in fact diverting the energy and funds of humane groups away from areas where animal abuse is common (such as the abandonment of 10 million pet dogs and cats each year to die of starvation, disease, and road accidents) in order to focus upon an area in which it is rare.[75]

The fact is that there *is* a growing spirit of self-criticism within the scientific community, but it is countered by an equally strong defensiveness, which is also on the rise. During the latest period of our survey (1982–1985), for every paper suggesting that improvements need to be made, another paper was written defending the status quo.

To consider in more detail how the topics described earlier have been fleshed out in the scientific literature, and how critical and defensive postures have taken shape in concrete terms, we will now turn to a discussion of specific themes.

Themes

Many of the themes considered in this section have already been introduced in the foregoing discussions of topics and tone. Here, however, we approach them more systematically. Recurring themes were organized into eight general subject areas, and the subject areas were further subdivided into relevant attitudes or opinions. On the general subject of the need for animals in research, for example, we found authors who (1) stressed the importance of animals in research, sometimes also (2) dismissing the potential of alternative methods; and others who (3) stressed the importance of alternatives. (As repeatedly stated, almost no one dismissed altogether the importance of animals in research.) Since (1), (2), and (3) are not mutually exclusive, some articles were found to contain more than one of these themes. Likewise, many items contain themes concerning more than one subject, or none at all.

The tabulation of the number of items in which each theme was expressed is presented in Table 5.3. To be consistent with previous presentations of data and to clarify trends over time, we have divided the 20-year span of the survey into 4-year periods. The classification of each theme, given in the column furthest to the right, represents an overall assessment of the importance of that theme in the literature. A *major* theme is one found in at least 15% of all 717 items that were read and coded. (Fifteen percent may not seem much, but one must take into account that many items—news briefs, for instance—contained no themes at all, yet were still included in the base. The percentage given for any theme is best evaluated in relation to the percentages for other themes rather than in absolute terms.) A *moderate* theme is one found in 8%–14% of all items, a *minor* theme in only 3%–7%, and a *negligible* theme in 2% or less of the total.

The following discussion of some of the highlights of the data presented in Table 5.3 is necessarily selective. A few of the themes are considered

TABLE 5.3. Themes expressed in articles on the ethical, humane, or regulatory aspects of animal experimentation published in the scientific literature in the United States, 1966–1985. The figures in each column represent the number of items (and percentage of items) in which the theme was found during each 4-year period.

	1966–1969	1970–1973	1974–1977	1978–1981	1982–1985	Total for 1966–1985	Classification[a]
(Total number of items)[b]	(83)	(56)	(81)	(153)	(344)	(717)	
1. On the need for animals in research and the feasibility of alternatives to the use of live animals							
a. Stressing the importance of animals in research, teaching, or testing.	13 (16%)	4 (7%)	12 (15%)	16 (10%)	64 (10%)	109 (15%)	Major
b. Dismissing the importance of alternative methods or their potential for success	1 (1%)	1 (2%)	4 (5%)	19 (12%)	31 (9%)	56 (8%)	Moderate
c. Stressing the importance of alternative methods and their potential for success	0	0	4 (5%)	28 (18%)	34 (10%)	66 (9%)	Moderate
2. On legislation and regulation							
a. Opposing further regulation; denial of any problem; animals are already treated humanely	18 (22%)	7 (13%)	11 (14%)	30 (20%)	64 (19%)	130 (18%)	Major
b. Advocating better enforcement for existing regulations	1 (1%)	0	4 (5%)	0	15 (4%)	20 (3%)	Minor
c. Advocating self-regulation through development of guidelines, accreditation of facilities, etc.	13 (16%)	1 (2%)	5 (6%)	23 (15%)	45 (13%)	87 (12%)	Moderate
d. Favoring more stringent regulations, including legislation	2 (2%)	1 (2%)	5 (6%)	6 (4%)	39 (11%)	53 (7%)	Minor
3. On animal welfare activists							
a. Hostile: Viewing them as a danger to research; characterizing them as antiscience, ignorant, emotional; news reports that emphasize antagonisms	11 (13%)	3 (5%)	8 (10%)	17 (11%)	59 (17%)	98 (14%)	Moderate
b. Conciliatory: Stressing common goals; news reports that emphasize rapprochement	0	2 (4%)	10 (12%)	20 (13%)	42 (12%)	74 (10%)	Moderate

continued

TABLE 5.3. (*continued*)

	1966–1969	1970–1973	1974–1977	1978–1981	1982–1985	Total for 1966–1985	Classification[a]
4. On the applicability to animals of ethical or moral considerations							
a. Dismissing ethical considerations as irrelevant to the use of laboratory animals; or treating the issue as a trivial one	3 (4%)	1 (2%)	7 (9%)	7 (5%)	6 (2%)	24 (3%)	Minor
b. Arguing that laboratory animals are entitled to moral or ethical considerations; or treating the issue as at least an important one for discussion	2 (2%)	1 (2%)	13 (16%)	39 (25%)	83 (24%)	138 (19%)	Major
5. Basis for scientists' interest in laboratory animal welfare							
a. Legalistic: Viewed primarily as a matter of following regulations	3 (4%)	3 (5%)	4 (5%)	6 (4%)	2 (1%)	18 (3%)	Minor
b. Technical or economic: viewed as a matter of ensuring valid scientific results; or not wasting valuable resources	4 (5%)	2 (4%)	3 (4%)	20 (13%)	25 (7%)	54 (8%)	Moderate
c. Empathetic: Viewed as a matter of alleviating pain and suffering	6 (7%)	4 (7%)	18 (22%)	25 (16%)	63 (18%)	116 (16%)	Major
d. Philosophical: Viewed as a matter of animal rights or human obligation, based on moral philosophy	0	0	4 (5%)	19 (12%)	50 (15%)	73 (10%)	Moderate
e. Religious or spiritual: Based on biblical or other religious authority; or on the "sanctity of life"	2 (2%)	1 (2%)	0	2 (1%)	3 (1%)	8 (1%)	Negligible

continued

TABLE 5.3. (*continued*)

	1966–1969	1970–1973	1974–1977	1978–1981	1982–1985	Total for 1966–1985	Classification[a]
6. Justifications for animal research							
a. Medical progress; benefits to humans in alleviating suffering	10 (12%)	0	13 (16%)	14 (9%)	70 (20%)	107 (15%)	Major
b. Knowledge for its own sake; scientific progress	4 (5%)	1 (2%)	2 (3%)	3 (2%)	13 (4%)	23 (3%)	Minor
c. Benefits to other animals	2 (2%)	0	5 (6%)	2 (1%)	21 (6%)	30 (4%)	Minor
d. The relatively greater value of human over animal life (viewing the issue as one of humans vs animals)	2 (2%)	1 (1%)	3 (4%)	2 (1%)	8 (2%)	16 (2%)	Negligible
e. "Waste" or "purpose" arguments: Most of the animals would have been killed anyway in pounds; or purpose-bred animals would not have lived at all	0	0	0	0	14 (4%)	14 (2%)	Negligible
f. Arguing that most research animals do not suffer	1 (1%)	0	2 (3%)	2 (1%)	12 (3%)	17 (2%)	Negligible
g. Presenting the issue as a matter of scientific freedom	4 (5%)	3 (5%)	2 (3%)	3 (2%)	15 (4%)	27 (4%)	Minor
h. Biblical or other religious grounds	1 (1%)	0	1 (1%)	4 (3%)	1 (0%)	7 (1%)	Negligible

continued

TABLE 5.3. (*continued*)

	1966–1969	1970–1973	1974–1977	1978–1981	1982–1985	Total for 1966–1985	Classification[a]
7. On the burden of proof							
a. The burden of proof lies on the investigator to justify the need for animals in each experiment; a cost-benefit analysis should be carried out in each case	0	1 (2%)	4 (5%)	9 (6%)	19 (6%)	33 (5%)	Minor
b. The investigator should not be burdened with having to justify the need for animals in each experiment; cost–benefit analyses are unworkable; one never knows which experiments will produce significant results	2 (2%)	1 (2%)	0	3 (2%)	0	6 (1%)	Negligible
8. On the need for education							
a. Stressing the need for educating the public about benefits of animal experimentation and/or the ethical uprightness of scientists	4 (5%)	0	1 (2%)	4 (3%)	28 (8%)	37 (5%)	Minor
b. Stressing the need for educating scientists, technicians, or students about humane care and ethical issues	4 (5%)	1 (2%)	2 (2%)	3 (2%)	11 (3%)	21 (3%)	Minor

[a]The classification scheme is based on the total percentage of items (1966–85) containing each theme: Major, 15% and above; Moderate, 8%–14%; Minor, 3%–7%; Negligible, 2% and below.

[b]Percentages are based on the number of items given in parentheses at the top of each column. The total is larger than the number of items included in the quantitative analysis because articles in books are here counted individually, whereas in the quantitative analysis each volume of papers published as a book was counted only once.

here at length; others are merely listed in the table with no further comment. We suggest that readers look especially at the themes numbered 5, 6, and 7 for additional data not discussed in the text.

The argument that animals are essential to scientific research forms a strong thread extending throughout the 20-year period of the study (Table 5.3, 1a), and it is almost always linked to a justification of such research in terms of its medical benefits (6a). Both represent major themes in the scientific literature. In 1968, for example, the American Society for Artificial Internal Organs (ASAIO) published a statement on the use of animals for research that began with the affirmation that "animal experimentation is an irreplaceable part of research and development of artificial organs."[76] A resolution adopted by the American Association for the History of Medicine and published in its *Bulletin* in 1966 stated that "past and future progress in medicine is entirely dependent on the continued use of animal experiments."[77]

This idea was expressed virtually unchanged 15 years later in a policy statement adopted by the American Pharmaceutical Association. The statement's first point was that the association "recognized that animal experiments continue to be an essential, and indeed irreplaceable, component of biomedical research and testing."[78] Even more recently, the American Veterinary Medical Association (AVMA) proposed a revision of its 1982 statement, the "Use and Regulation of Research Animals"; in its report to members, the AVMA reassured them that "as before, the statement recognized the essential role of animals in research, experimentation and testing."[79]

Occasional opinions to the contrary have been voiced within the past few years in letters to editors, but not in position papers carrying the weight of a professional organization. A few dissidents have questioned the need for continued animal research in general,[80] or denied the need for animals in specific areas.[81] On the whole, however, we found little controversy about the importance of animals to research, teaching, or testing. One report from the literature even provides some outside support for this observation. Respondents' comments in a survey on animal use in teaching conducted by the Association of Chairmen of Departments of Physiology "range[d] from a minority view that animal laboratories are unnecessary and can be replaced by computer models to a majority emphasizing that they are irreplaceable for the student's learning experience." The author noted wryly, "There are unfortunately no objective criteria available to validate either viewpoint."[82]

The value of animals to research has not usually been denied, even by the most optimistic advocates of alternative methods. We found no claims in the scientific literature that animals can be replaced altogether in the laboratory within the foreseeable future. (This is not to say that such claims have not been made by others, especially in the animal rights literature.) The pro-alternatives viewpoint at its strongest in the scientific literature is represented by the following statement made by Andrew Rowan:

> The concept of alternatives will not, if enthusiastically applied, result in the immediate cessation of animal experimentation. In many areas of research a problem can only be investigated by means of an intact animal. On the other hand, the number of animals being used in research today could be substantially reduced if more effort and attention were given to promoting the idea of alternatives. At the same time, there need be no threat to the quality of biomedical research or to the advance of biological and medical knowledge.[83]

A whole chapter in the volume cited above is devoted to advances in biomedical knowledge and improved therapies made possible through animal experimentation, even though the rest of the book makes it clear that Rowan has no "wish to add [his] voice to the general paean of praise for animal research per se."[84]

Unlike the near consensus on the importance of animals in (at least some) biomedical research, the question of alternative methods has produced a lively debate, starting at the end of the 1970s. At first, the optimists held a slight edge: During 1978–1981 we found 28 favorable opinions to 19 unfavorable (Table 3.5, 1b and 1c). From 1982 through 1985, pros and cons were found in roughly equal proportions.

Not all of those who stressed the importance of alternatives did so with quite Rowan's vigor, but many displayed the kind of support for the idea expressed by two Rockefeller University researchers who phrased the issue as follows: "Is the necessary knowledge and technology available to replace certain *in vivo* tests? We respond to this question in the affirmative, with reasonable optimism."[85] It is no longer unusual, either, to find scientists raising the kinds of questions asked by two veterinarians in a letter to *JAVMA*:

> Are there not serious ethical questions raised regarding eye irritancy testing of human cosmetic products on animals? Shouldn't we as a profession do all we can to replace the animals in these laboratories with nonanimal testing methods?[86]

Still, in the most recent years, we found just about as many others on the side of the author who called alternatives "a nonissue if ever one existed."[87] Those taking the negative side in this particular debate often dismiss the potential for alternatives in order to bolster their arguments for the necessity of animals in research. A spokesman for the American Physiological Society used this tactic when he testified at a Congressional hearing that "the need for laboratory animals in research is crucial and will continue inasmuch as there are no nonanimal adjunct methods available that can replace them."[88] In a similar vein, an official of the National Institutes of Health (NIH) wrote a letter to *Science* emphasizing improvements in medical care made possible by animal research, to which he added, "The only *real* alternative to animals in the laboratory is a loss in the rapid gains being made in improving health."[89] Two psychologists cited earlier for their defensive posture wrote a detailed assessment of the potential for alternatives in psychological research and concluded (in 1985) that "at the present time, there are no viable alternatives to the use of live organisms in behavioral research."[90]

The debate over alternatives, like all other aspects of the controversy, usually boils down to a debate about legislation. As part of the intense legislative activity of the 1980s, several bills were introduced that would have provided support for the development of alternative methods, along with restrictions on animal use. In an editorial opposing one of these bills, a spokesperson for the Association of American Medical Colleges (AAMC) incorporated most of the defensive themes listed in Table 5.3. He wrote that animals are necessary to further progress in medical treatment and that alternative methods are not adequate. He added the frequently heard argument that scientists already treat animals well because abuse would compromise the data. As he put it, "The fact that well nourished, healthy animals are essential for obtaining valid data in research requires that they be properly cared for and humanely tested." In short, "The AAMC believes that standards established by the Animal Welfare Act of 1966 are sufficient to achieve the desired goal."[91] Nevertheless, as we have pointed out, three laws passed in 1985 did ultimately give legislative backing to the concept of alternatives.

Regulatory concerns, then, are consistently the number one topic. Table 5.3 (2a) shows, in addition, that most scientists expressing opinions on the subject are against further regulation of any sort; for them, there is no problem that needs correction. Often they do profess an interest in animal welfare, but like the AAMC spokesperson quoted above, they believe that the need to obtain valid scientific results and to conserve valuable resources is sufficient to ensure humane treatment of laboratory animals (Table 5.3, 5b).

The same arguments, in fact, had been used to oppose the 1966 act. According to an article published that year, the public did not realize that

> the vast majority of laboratory animals receive much better care than farm animals or for that matter household pets. It does not make good sense to treat laboratory animals otherwise since research protocols call for healthy models in order to reduce experimental variables.... Only in rare instances are animals exposed to painful or traumatic experiences without prior induction with appropriate local or general anesthetic agents.[92]

This author thought that everything was so rosy in animal laboratories that he opposed even a system of self-regulation. A scientist's time, he wrote, is better spent in "the merciful work of alleviating suffering and prolonging life."

Echoes of that attitude have reappeared regularly. In a report on the testimony given by scientists at a Congressional hearing in 1981, the *Physiologist*'s columnist wrote that all seven of the scientists had opposed the proposed legislation:

The researchers also noted that there is very little misuse of animals in research inasmuch as it is in the best interest of the scientist to treat laboratory animals humanely to obtain valid results. It was stated that whatever the extent of abuse of animals in the laboratory, it is far less than the number of animals killed or mistreated in cities and on farms.[93]

In a similar vein, a veterinarian observed that, in 30 years of research work, he had never seen "wanton cruelty." He argued that no changes are needed in the way things are done because "the scientific community is strongly committed to humane treatment of animals not only because it is an ethical imperative, but also because it is an important control factor in research."[94] Likewise, a spokesperson for the American Physiological Society in 1984 argued that

> by and large, statements of malicious laboratory animal abuse by those who oppose animal research and scientific inquiry are anecdotal and are largely without verification. Such statements are designed, in my opinion, to sow seeds of mistrust about science and research in the minds of those unfamiliar with the standards and regulations that govern all research practices involving the use of laboratory animals.
>
> It is because of such anecdotal testimony that the Society maintains its position that restrictive legislation is not needed.[95]

For people who believe that nothing is wrong, the current pressures to make improvements are incomprehensible. "Such criticism tends to confound us," wrote two researchers.[96] A physiologist also seemed confounded when he wrote:

> It is difficult for experimental physiologists to understand what the flack is all about in reading newspaper editorials and watching television accounts of alleged cruelty or mistreatment of animals in a modern research laboratory. In our real world we know such cruelty or mistreatment to be very rare....[97]

In spite of the prevailing opposition to legislation, attitudes have shifted slightly over the years. A speaker who addressed the American Veterinary Medical Association in 1975 commented that by that time at least the Animal Welfare Act had become accepted. Before 1966, she said,

> The idea of outside "interference" or "policing" was anathema to most experimenters—or perhaps, more accurately, to those whose views on the subject could be heard. I add this qualification because in the early 1960s, it was essentially impossible for a scientist to have a letter favoring outside regulation published in any scientific journal.[98]

This scientist did succeed in getting such a letter published in *Science* a couple of years later,[99] and we found dozens of examples of support for some kind of outside regulation that appeared in the literature from 1982 through 1985 (Table 5.3, 2d). Unusual enthusiasm for legislation was displayed in a 1981 article written by a veterinarian who helped draft an amendment to the Animal Welfare Act that was first introduced in

Congress the previous year. He wrote that it is "very naive" to think that all scientists follow the NIH guidelines, and asked his colleagues:

> If we do believe that we would serve the scientific community, the animals, and society better by collectively coercing sound care and use of laboratory animals across the board, ought we not be those who desire a comprehensive animal welfare law the most?[100]

Even among supporters of legislation, such wholehearted enthusiasm remains rare. Most of those who expressed support for legislation did so in the spirit of the letter writer who warned scientists that if they do not do a better job of regulating themselves, legislation is warranted.[101]

Still, it is indicative of a change in the air to note that the proportion of opinions favoring more stringent regulation was up during the most recent period, from 4% to 11% (Table 5.3, 2d). This reflects some support for the Animal Welfare Act of 1985 and the new Public Health Service (PHS) directives, even among members of a scientific community generally opposed to greater restrictions. The bill sponsored by Douglas Walgren in the House of Representatives that was eventually incorporated into the PHS Health Research Extension Act (PL 99–158) was supported by the American Veterinary Medical Association.[102] The bill sponsored by Robert Dole in the Senate that was passed as an amendment to the Animal Welfare Act was also supported by the AVMA,[103] as well as by the American Physiological Society (APS).[104] The APS was not happy about endorsing the legislation, but gave it "qualified support" as the lesser of evils, noting at the same time that "the Society maintains its position that restrictive legislation is not needed."[105] When the Dole amendment was finally passed by the Senate in 1985, the APS took credit for "provid[ing] the concept to Dole" of writing the bill as an amendment to the Animal Welfare Act (which would affect virtually all research facilities), rather than as a measure that would affect only federally funded institutions.[106]

The biggest change, however, is the introduction of ethics into the debate. Earlier we demonstrated the sudden appearance of ethical topics midway through what we have called this "era of transition." Once ethics had been introduced as a legitimate concern, few people actually came out in opposition. In fact, after being a nontopic for years, the argument that laboratory animals are entitled to ethical or moral considerations became a major theme during the past decade (Table 5.3, 4b).

The application of ethical precepts to animals is a more controversial issue than it appears, however. There may be something close to a consensus on such vague statements as the pharmacologist's view quoted earlier that "ethics can contribute" to animal research, but even that is sometimes challenged. Nobel Prize winner David Baltimore, a microbiologist, declared that he sees no moral issue involved in animal research.[107] A similar statement was made by two psychologists, who wrote that "the evolution of moral and ethical behavior in man may be such that it is not applicable to other species."[108] In the opinion of a veterinarian, "domestic animals exist in this world because they fulfill man's needs ... therefore it is meaningless to speak of their rights to existence, because they would not exist if man did not exist."[109] For these people, "The dispute is not over a moral issue, but over whether there is a moral issue," as *MD* put it as recently as 1981.[110] These kinds of opinions were found in only 3% of all the items, however, so they constitute a very minor—almost negligible—theme.

We could point to a far greater number of examples from the recent literature urging that ethics be given a larger role in scientific research with animals,[111] that ethical considerations be included in training scientists,[112] that there are serious ethical dilemmas involved in animal research,[113] or referring to the "ethical questions" or "moral problems" raised by animal research.[114] Some go so far as to ask, "What morally relevant differences are there between human and nonhuman animals...?"[115] This kind of talk was not part of the discourse of the 1966 debates over the Animal Welfare Act.

When it comes to deciding how to apply ethics to animal research, even the appearance of consensus melts away. The letters columns of *Science* and *JAVMA* are full of debates over what sort of regulation is needed, what kinds of research are justifiable, under what circumstances animals can be ethically used in teaching, whether pounds should be required to release animals to research institutions, and what sorts of rights animals have.

A debate in *JAVMA* over this last issue provides an excellent illustration of the ferment that has begun to bubble in the literature. Early in 1982, the newly consituted Animal Welfare Committee of the American Veterinary Medical Association (AVMA) adopted a set of Guiding Principles, which were published in *JAVMA*. The principles stated, in part, that "AVMA positions should be concerned primarily with the scientific aspects of the medical well-being of animals rather than with the philosophic or moral aspects." Furthermore, "The AVMA believes that the use of the term 'animal rights' has to do with personal philosophical values and therefore recommends that the term 'animal rights' not be used, and encourages the profession to focus its attention on the welfare and humane treatment of animals."[116]

This statement elicited a stream of letters, both supporting and disapproving, that continued into 1984. The dispute was periodically refueled whenever the committee published another report reaffirming, and eventually modifying, its position. We counted 11 letters directly responding to the AVMA position on animal rights, and another 13 on the subject of animal rights that were published during the same period of time. The letters overwhelmingly supported the thesis that animals have rights (14 to 4, with 6 somewhere in between). Many criticized the AVMA for not endorsing the concept of animal rights, which one veterinarian called "one of the backbones of our profession."[117] By early 1984, the AVMA position had budged slightly, to add to the previous statement the assertion that "it is also recognized that veterinarians have certain ethical, philosophical, and moral values that must be considered."[118]

The place of ethical concerns in animal experimentation has been transformed during the era under discussion. There is little agreement about what "ethical concerns" are or how they should be applied, but the concept is now secure, at least as a starting point. Another illustration of this shift comes from comparing the first report on euthanasia prepared in 1963 by the American Veterinary Medical Association with the latest one published in 1986. The 1963 report, which did not refer to laboratory animals, contained the following sentence:

> The ethical and professional problems involved in electing to perform euthanasia upon an animal were not considered to be within the scope of the Panel's activities.[119]

The 1986 report, which puts considerable emphasis on laboratory animals, replaced the above sentence with the following:

> The ethical and psychologic issues involved with euthanasia of diseased, injured or unwanted animals is within the purview of this discussion. Moral and ethical imperatives associated with individual or mass animal euthanasia should be consistent with acceptable humane practice. In all circumstances, choice of method should be selected and employed with the highest ethical standards and social conscience.[120]

This recent acceptance of the legitimacy of ethical concerns, along with a growing openness to alternatives among some segments of the scientific community, has been accompanied by a change in the way animal welfare activists are portrayed. A letter published in *Science* in 1966 contained a classic statement of what was probably then the dominant stereotype:

> Regardless of the category of "humanitarianism" from which it issues, the constant barrage of attacks, some of them scurrilous, is in the greatest degree anti-intellectual and in the know-nothing tradition. It is not coincidence that from time to time we find these "humanitarians" embracing other fringe causes. The little old lady in sneakers wears a mink, rides a chauffeur-driven Rolls Royce, and supports a Cause. These people probably distrust any intellectual effort in any field.[121]

Animal welfare organizations have been the traditional enemy of experimental science since the 19th century, and expressions of enmity and condescension toward them are still frequent in the scientific literature. Before 1974 a kind word for animal welfare activists was scarce indeed (only 2 out of 139 items) (Table 5.3, 3b). One was more likely to find opinions like the statement in another 1966 article that animal welfare people suffer from "a disease condition, which has been designated by the name, zoophilic psychosis, in which there is an inordinate and exaggerated sympathy for the lower animals often associated with delusions that they are persecuted by man...."[122]

Since then, the countertheme of rapprochement with animal protectionists has become more popular (Table 5.3, 3b). A 1972 *JAVMA* article entitled "Trends in Humane Animal Care Legislation" opened with the comment that "a title such as this, in the minds of many of us, conjures up the image of groups of 'little old ladies,' complete with tennis shoes and floppy hats." The author went on to say that this image was now "only partly true" and should not prevent veterinarians from taking animal welfare issues seriously.[123]

A few years later, a United States Department of Agriculture spokesperson declared in *JAVMA*'s pages that "we must be ready to accept or at least consider important contributions from other biomedical disciplines and animal welfare organizations."[124] A paper delivered at the 1980 convention of the American Psychological Association emphasized the "constructive elements" of the animal protection movement.[125] An editorial in *MD* in 1981 claimed that "the people now behind the animal-rights' controversy have earned the right to be taken very seriously indeed."[126] That same year, *Lab Animal* took the unprecedented step of publishing an interview with animal rights activist Henry Spira, who also took a conciliatory approach in stressing that "there is no necessary opposition between persons concerned with the rights of animals and persons concerned with scientific and intellectual pursuits. I ... hope that, rather than a clash of concerns, we can find common ground."[127] By 1982 *Science* was reporting that "leaving out the extremists on both sides of the question, scientists and animal welfare people do not apppear to be much in conflict."[128] Two years later, we found numerous scientists recognizing animal welfare groups for raising their consciousness and sensitivity,[129] and predicting that "there are signs that the parties in the laboratory animal welfare controversy may be ready to recognize each other's concerns."[130]

On the other hand, many scientists still do consider themselves very much in conflict with animal welfare people. Table 5.3 (3a and 3b) shows that from 1982 through 1985 there were more hostile than conciliatory attitudes expressed. Conciliatory attitudes held a slight majority during the preceding period (1978–1981), so this current trend represents a shift in the balance of (published) opinion. It is quite clear from our data that, in spite of the increasing number of conciliatory attitudes expressed, they are not increasing relative to those that see only a threat in the concerns of animal protectionists. Typical of the latter is a prominent psychologist quoted in a 1982 item in *Science*, who said that "he and his colleagues are 'terribly concerned' about the animal rights movement and believe it poses 'considerable danger to the freedom of research,' particularly basic behavioral and physiological research."[131]

Much of the hostility, now as well as in earlier times, stems from a belief that animal welfare organizations would like to put a stop to animal research altogether. Thus, the director of the American College of Surgeons wrote in a 1982 editorial that "recent legislative bills have deceptive features that obscure their basic anti-science and antivivisectionist thrust. It is in the emphasis on 'alternative methods' of research that such bills hide the end result being sought." Instead of cooperation, the editorial calls for "resistance by the scientific community ... and vigilance organized on a full-time basis."[132] Even more recently, a columnist warned readers that "there should be no doubt that the ultimate objective of the animal protection groups is the abolition of animal research."[133]

Convincing legislators that animal research should not be abolished is therefore seen as the priority issue by many scientists who enter the debate and by virtually all of those who view animal protection groups as a threat to research. Persuading legislators of anything generally means gaining the support of their constituents, so these scientists stress the need to educate the public about the benefits of animal experimentation and the high standards of humane care that they claim are already in effect. For example, the columnist quoted above went on to urge the scientific community to "explain itself and its methods to the public, the ultimate arbiter of public policy."[134] We found this theme to be of minor importance in the literature (Table 5.3, 8a), but that counts only the explicit suggestions for public education campaigns. In addition, the theme is implied in every justification of animal research and every article extolling its benefits for human welfare. Moreover, explicit statements like the following were found more often during 1982–1985 than at any previous time:

> It remains for the scientific community to convince the general public that research on animal subjects must continue and is vital to progress in biomedical research. Furthermore, that the animals are treated humanely and not used unnecessarily. If we as scientists are to succeed, an unremitting educational effort on our part will be required. Because the necessity of animal research is so obvious to those of us in science, we forget that the general public is increasingly removed from exposure to the scourges of famine and infectious disease and is less and less aware of the essential role animal research continues to play in the conquest of these challenges to civilized existence.[135]

The assumption contained in that statement and in many similar ones[136] is that criticism of animal research practices is based on ignorance. Although that assumption may well be accurate regarding some of the more extreme antivivisectionist claims, it misses the mark when used as a blanket defense, especially against the criticisms raised by other scientists. This observation highlights an odd quality to the debate: When critical scientists find fault with various animal research practices, they often find the response directed, not to them, but to a presumably misguided and uninformed public.

With the passage of the 1985 legislation and revisions in the PHS Policy, there is an opening now for more dialogue within the scientific community as researchers come to grips with the new regulations. In their fear that animal research would be drastically curtailed by the new law, most scientists did not feel secure enough to admit much of a need for their own education in ethical issues or the humane handling of animals. The need for education of researchers constitutes a very minor theme in the literature. From 1982 through 1985, 28 authors called for public education campaigns to gain support for their view of the value of animal research; less than half that number (11) suggested that educational programs be instituted for scientists, technicians, or students. The widespread assumption seems to be that researchers already know all they need to know about how to administer the most humane care to laboratory animals and how to recognize and deal with the ethical issues of animal research (Table 5.3, 8a and 8b).

The Scientists Center for Animal Welfare (SCAW), on the other hand, has taken a dimmer view of researchers' inherent proficiency in these matters. SCAW has been a leading proponent of training programs for scientists, as reflected by the large number of proposals along those lines suggested at a conference organized by SCAW in 1982.[137] One of the 12 recommendations coming out of that conference, representing a consensus of the participants, was that

> *training courses be provided to scientists to increase their empathy concerning the animals they are using, as well as knowledge about established animal care policies*: especially for review committee members; those using primates or endangered species; those using highly invasive techniques. (Emphasis in original)[138]

With the threat of the abolition of animal research at least temporarily removed, we expect that this theme will take on greater importance as scientists shift their attention to focus on ways to accomplish both good research and good ethics. In the following chapter, we discuss the need for an increased emphasis in the future on the education of scientists.

Summary

It was suggested in Chapter 4 that the growing proportion of news briefs in the research animal literature might indicate a narrowing of the focus of the debate as the 1985 legislation neared passage. Data presented in this chapter showed that the proportion of items concerned with regulatory issues during the most recent period was indeed higher than at any time during the previous 8 years, but that it was much lower than during the first decade covered by this study (Table 5.1). Prior to 1974 the scientific community approached the controversy almost purely as a matter of rules and regulations. Since then, the debate has in fact *enlarged* its focus. Ethical issues, practical approaches to animal pain, other forms of distress, and suffering, and the development of alternatives to the use of animals in research are topics that have gained attention since 1974. Although, in a sense, these topics are inextricable from regulatory matters, they have all gained some measure of attention in their own right.

The most recent period, from 1982 through 1985, was not, however, the high point of concern for any of these topics. A concentration on regulatory issues peaked in the early 1970s; on pain, distress, and suffering, in the late 1970s; and on ethics and alternatives, in the early 1980s. The most recent trend is a reversion to earlier patterns, with regulatory issues once again gaining ground. Given scientists' preoccupation with legislative developments that culminated in 1985 in the passage of the Animal Welfare Act amendments, the Health Research Extension Act, and the new PHS Policy, it is perhaps surprising only that the pendulum swung back no farther than it did.

As this era of transition in animal research draws to a close, we find a near consensus on the value of animals to research and on the need to apply ethical concepts to animal experimentation, but a real division on most specifics and, indeed, on the need for any change. This division has surfaced over and over again: in the growing numbers of scientists who are speaking out to criticize current practices, along with the growing numbers who are more concerned with justifying the status quo; in the greater sense of common cause with animal protectionists felt by many researchers, along with greater hostility to animal advocates expressed by others; in the arguments over the potential for alternatives; and in attitudes toward the new regulations.

In sum, the scientific community faces the coming years less solidly united on many issues than it was entering the post-Animal Welfare Act era in the late 1960s. The controversy is obviously far from finished. A *Newsweek* cover story in late 1988 noted growing public support for the animal rights movement and concluded that "scientists can no longer afford to pretend that their critics' moral concerns are frivolous. Profound questions are being raised, and ignoring them won't make them go away."[139] We take it as an encouraging sign that the scope of the discussion in its current phase involves a larger segment of the scientific community and a broader range of issues than did the debate over the first federal legislation affecting animal research. These changes may make possible a more constructive dialogue among disparate elements in the research community than has been the case in the past.

Notes

(The notation *[AB]* after a reference indicates that it is included in the Annotated Bibliography following Chapter 7.)

1 Examples of feature articles include the following: Langer E (1966) Animal care: The humane movement is pulling ahead. *Science 151*:1515–1518 [AB]; Visscher MB (1971) The Animal Welfare Act of 1970 (P.L. 91–579) *Science*

172:916–917 [AB]; Holden C (1974) Beagles: Army under attack for research at Edgewood. *Science 185*:130–131 [AB]; Smith RJ (1977) Creative penmanship in animal testing prompts FDA controls. *Science 198*:1227–1229 [AB]; Broad WJ (1980) Legislating an end to animals in the lab. *Science 208*:575–576 [AB]; Holden C (1981a) Police seize primates at NIH-funded lab. *Science 214*:32–33 [AB]; Holden C (1981b) Scientist convicted for monkey neglect. *Science 214*:1218–1220 [AB]; Fox JL (1984a) Lab animal welfare issue gathers momentum. *Science 223*:468–469 [AB]; Fox JL (1984b) Changes in animal care policy proposed. *Science 224*:364–365 [AB]; Fox JL (1984c) Animal rights bill defeated in California. *Science 224*:1414 [AB]; Fox JL (1984d) Rifkin takes aim at USDA animal research. *Science 226*:321 [AB]; Culliton BJ (1985) HHS halts animal experiment. *Science 229*:447–448 [AB].

[2] For example, Leeper EM (1976) NAS symposium: New methods may reduce needs, can't replace laboratory animals. *BioSci 26*,1:9–12 [AB]; Eskridge NK (1978) Researchers concerned about changes in animal care policy. *BioSci 28*,6:409–410 [AB]; Henig RM (1979a) Stress science, not ethics—animal experimentation: The battle lines soften. *BioSci 29*,3:145–148,195–196 [AB]; Henig RM (1979b) Animal welfare groups press for limits in high school research. *BioSci 29*,11:651–653 [AB]; Walton S (1981) Congress puts bioassay to the test. *BioSci 31*,4:287–290 [AB]; Wortman J (1985) PHS revises its lab animal welfare policy. *BioSci 35*,7:408 [AB].

[3] The column was inaugurated with Samuels W (1981) Congressional hearings on the use of animals in medical research and testing. *Physiol 24*,6:13–14 [AB].

[4] For example, Trull FL, Kalikow NA (1984) Animal rights movement: A threat to biomedical research? *Cancer Investig 2*,6:479–482 [AB].

[5] Eccles JC (1971) Animal experimentation versus human experimentation. In: National Research Council-Institute of Laboratory Animal Resources *Defining the laboratory animal.* Washington, DC: National Academy of Sciences, pp 285–293 [AB].

[6] Simpson LL (1983) Animals and animal tissues in the discipline of pharmacology. In: Sechzer JA (ed) *The role of animals in biomedical research.* Annals of the New York Academy of Sciences, vol 406. New York: The New York Academy of Sciences, pp 74–81 [AB].

[7] Hirschboeck JS (1966) Animals and human welfare. *Hosp Progr 47*,5:98–99,122 [AB].

[8] Ingle DJ (1966) Scientific and ethical responsibilities associated with the use and care of animals. *Fed Proc 25*,5:1465–1467 [AB]; Roberts C (1967) *The scientific conscience: Reflections on the modern biologist and humanism.* New York: George Braziller, Inc. (1974, Fontwell, Sussex: Centaur Press) [AB].

[9] Ladimer I (1977) Root and branch: Legal aspects of biomedical studies in man and other animals. In: National Research Council-National Academy of Sciences, Institute of Laboratory Animal Resources *The future of animals, cells, models, and systems in research, development, education and testing.* Washington, DC: National Academy of Sciences, pp 296–314 [AB].

[10] Brown RE (1983) Rights of animals. [letter] *J Am Vet Med Assoc 182*,5:444–445.

[11] Fox MW (1983a) Use of animals for teaching purposes. [letter] *J Am Vet Med Assoc 182*,12:1314–1315.

12 *Pain*, the journal of the International Association for the Study of Pain, was not included in this survey because it is not published in the United States. Had it been included, we could have cited many more articles on animal pain.

13 PL 91–579 (Dec 24, 1970) 84 Stat 1560–65, Sec 14 (to amend Sec 13 of the Animal Welfare Act).

14 New directives for Animal Welfare Act. (1975) *J Am Vet Med Assoc 167*,4:260.

15 Melby EC Jr (1976) What does "adequate veterinary care" mean to the American College of Laboratory Animal Medicine? *J Am Vet Med Assoc 168*,6:508–509 [AB]; Vierck CJ Jr (1976) Extrapolations from the pain research literature to problems of adequate veterinary care. *J Am Vet Med Assoc 168*,6:510–513 [AB]; Anchel M (1976) Beyond "adequate veterinary care." *J Am Vet Med Assoc 168*,6:513–517 [AB]; Schwindaman DF (1976) An evaluation of adequate veterinary care by the US Department of Agriculture. *J Am Vet Med Assoc 168*,6:517–519 [AB].

16 Dawkins MS (1980) *Animal suffering: The science of animal welfare*. New York: Methuen, Inc (also published in London by Chapman and Hall) [AB].

17 Kitchell RL, Erickson HH, Carstens E, Davis LE (eds) (1983) *Animal pain: Perception and alleviation*. Bethesda, MD: American Physiological Society [AB].

18 AVMA colloquium on recognition and alleviation of animal pain and distress. (1987) *J Am Vet Med Assoc 191*,10:1186–1298.

19 Ingle DJ (1966) Op cit [note 8].

20 Antivivisection rides again. (1971) *J Am Med Assoc 217*,1:70 [AB].

21 Pratt D (1976) *Painful experiments on animals*. New York: Argus Archives [AB].

22 Pratt D (1980) *Alternatives to pain in experiments on animals*. New York: Argus Archives [AB].

23 Leeper EM (1976) Op cit [note 2].

24 Bustad LK, Hegreberg GA, Padgett GA (1977) Animal models. In: NRC-NAS, ILAR, op cit [note 9], pp 130–149 [AB].

25 Rack L (1981) Bookshelf: *The principles of humane experimental technique* and *The use of alternatives in drug research*. *Lab Anim* (NY) *10*,4:57–58 [AB].

26 Rowan AN (1981) The test tube alternative. *The Sciences 21*:16–19,34 [AB].

27 For news reports and analysis of the bills, see Rowan AN (1980) The new federal bills: What they are and what they mean for animal research. *Lab Anim* (NY) *9*,4:25–28 [AB]; Broad WJ (1980) Op cit [note 1]; Walton S (1981) Op cit [note 2].

28 Harriton L (1981) Conversation with Henry Spira: Draize test activist. *Lab Anim* (NY) *10*,1:16–22 [AB].

29 Gunby P (1981a) Animal rights group awards research grant. *J Am Med Assoc 245*,18:1803; Gunby P (1981b) Search for alternatives to animal testing expands. *J Am Med Assoc 246*,18:2013–2014; Smith RJ (1981) Revlon funds animal test research. *Science 211*:260.

30 Walton S (1981) Op cit [note 2].

31 Samuels WM (1985) People you should know ... Doug Walgren. *Physiol 28*,1:5–7 [AB].

32 GAO finds flaws in Department of Agriculture performance on animal welfare. (1985) *J Am Vet Med Assoc 187*,7:694–695.

33 Fox MW (1984) Reconciling humane concerns and laboratory animal research. [letter] *N Engl J Med 310*,5:325–326.

[34] Rowan AN (1984) *Of mice, models, and men: A critical evaluation of animal research.* Albany, NY: State University of New York Press [AB].
[35] Scientists Center for Animal Welfare Newsletter (1984) Canadian view on animal welfare. [reprint] *J Am Vet Med Assoc 184*,7:792.
[36] Pratt D (1976) Op cit [note 21]; Pratt D (1980) Op cit [note 22].
[37] Federation of American Scientists (1977) Animal rights. *Pub Interest Rep 30*,8:1–8, as quoted in Eskridge NK (1978) Op cit [note 2].
[38] Holden C (1978) New group designed to draw scientists to animal cause. *Science 202*:35.
[39] Loew FM (1978) Animal rights in the laboratory. [letter] *Science 201*:482,484.
[40] Silverman P (1978) *Animal behavior in the laboratory.* New York: Pica Press [AB].
[41] Carter L (1979) NEI votes to protect cold-blooded animals. *Science 206*:1383.
[42] Rollin BE (1985) The moral status of research animals in psychology. *Am Psychol 40*,8:920–926 [AB].
[43] Rollin BE (1978) Updating veterinary medical ethics. *J Am Vet Med Assoc 173*,8:1015–1018 [AB].
[44] Caplan AL (1983) Beastly conduct: Ethical issues in animal experimentation. In: Sechzer JA (ed) Op cit [note 6], pp 159–169 [AB].
[45] Harriton L (1981) Op cit [note 28].
[46] Stevens CG (1977) Humane perspectives. In: NRC-NAS, ILAR, op cit [note 9], pp 16–24 [AB].
[47] McArdle JE (1984) Animal rights movement. [letter] *Science 225*:114–116.
[48] Seiling E (1980) Animals in the lab. [letter] *Science 209*:214.
[49] Rowan AN (1981) Op cit [note 26].
[50] Rowan AN (1984) Op cit [note 34]; see also Rowan AN (1980) Op cit [note 27]; Rowan AN (1980) [letter] *J Am Vet Med Assoc 176*,11:1208,1210; Rowan AN (1981) Can alternatives be integrated with animal models? *Lab Anim* (NY) *10*,2:38–45 [AB]; Rowan AN, Rollin BE (1983) Animal research—For and against: A philosophical, social, and historical perspective. *Perspec Biol & Med 27*,1:1–17 [AB]; Rowan AN (1984) The future of animals in research and training. The search for alternatives. *Fund Appl Toxicol 4*,4:508–516 [AB]; Rowan AN (1985) More on animal rights. [letter] *BioSci 35*,6:330; Rowan AN, Goldberg AM (1985) Perspectives on alternatives to current animal testing techniques in preclinical toxicology. *Ann Rev Pharm Toxicol 25*:225–247 [AB].
[51] Not to be confused with Michael A. Fox, a Canadian philosopher who until recently defended animal experimentation. His about-face on this issue startled the scientific community in 1986. For works expressing Fox's (now disavowed) defense of animal research, see Fox MA (1978) Animal liberation: A critique. *Ethics 88*:106–118; Fox MA (1980) On justifying the use of animals for human ends. *Anim Regul Stud 2*:191–204; Fox MA (1985) Ethical considerations in ethical animal research. *Adv Pain Res Ther 9*:685–694; Fox MA (1986) *The case for animal experimentation.* Berkeley: The University of California Press. The announcement of his reversal was published in a letter to *The Scientist* (Fox MA [1986] Author reverses views on animal rights. *The Scientist 1*[Dec 15]:10). (We thank D.S. Kronfeld, School of Veterinary Medicine, University of Pennsylvania, and Harry C. Rowsell, Canadian Council on Animal Care, both of whom independently supplied this reference.)

[52] Fox MW (1983a) Op cit [note 11]; Fox MW (1984) Op cit [note 33]; Fox MW (1978) Humane ethics in veterinary education. *J Vet Med Educ 5*,2:98–101 [AB]; Fox MW (1979) [letter] *J Am Vet Med Assoc 175*,4:329,342–343; Fox MW (1981) Experimental psychology, animal rights, welfare, and ethics. *Psychopharmacol Bull 17*,2:80–84 [AB]; Fox MW (1982) On the use of animals in research. [letter] *Am Psychol 37*,5:598–599; Fox MW (1983b) Changing attitudes on animal worth. [letter] *J Am Vet Med Assoc 183*,2:163,166; Fox MW (1983c) Equal and fair treatment for animals. [letter] *J Am Vet Med Assoc 183*,11:1133–34; Fox MW (1985) Who comes first, animals or people? [excerpt reprinted from *Intervet*, Oct 1984] *J Am Vet Med Assoc 186*,4:347; Kindler RA, Fox MW (1985) Animal rights and speciesism. [letter] *J Am Vet Med Assoc 186*,2:123–124.

[53] Fox MW (1985) Ibid.

[54] Fox MW (1981) Ibid.

[55] Bowd AD (1980a) Ethics and animal experimentation. *Am Psychol 35*:224–225 [AB]; Bowd AD (1980b) Ethical reservations about psychological research with animals. *Psychol Rec 30*,2:201–210 [AB].

[56] Keith DA, Donoff RB (1981) Use of animals questioned. [letter] *J Oral Surg 39*,3:169. A reply from the experimenters follows; Reid DA (1985) An animal model of myopia. [letter] *N Engl J Med 313*,18:1162–1163. A reply from the experimenters follows.

[57] Segal E (1982) Editorial. *J Exper Analysis Behav 38*:116 [AB].

[58] Wolff NC (1983) AVMA and animal welfare position. [letter] *J Am Vet Med Assoc 183*,1:36.

[59] Vivisection-vivistudy: The facts and the benefits to animal and human health. (1967) *Am J Pub Health 57*,9:1597–1626.

[60] Winterscheid LC (1967) Animal experimentation leading to the development of advanced surgical techniques. *Am J Pub Health 57*,9:1604–1612 [AB].

[61] McCann MB, Stare FJ (1967) The contribution of animal studies to nutritional discoveries that have benefited both animal and man. *Am J Pub Health 57*,9:1597–1604 [AB].

[62] Robinson HJ (1967) Animal experimentation leading to the development of drugs benefiting human beings and animals. *Am J Pub Health 57*,9:1613–1620 [AB].

[63] Soave OA (1967) Animal experimentation leading to better care of laboratory and pet animals. *Am J Pub Health 57*,9:1621–1626 [AB].

[64] Best CH (1974) A short essay on the importance of dogs in medical research. *Physiol 17*,4:437–440 [AB].

[65] Davison FC (1977) Historical perspectives of biomedical experimentation. In: NRC-NAS, ILAR, op cit [note 9], pp 7–15 [AB]; Pattishall EG Jr (1977) Animal behavior: Relation to illness and disease. In: Ibid, pp 115–123 [AB].

[66] Harrell GT (1977) Introduction. In: Ibid, pp 1–6 [AB].

[67] Migaki G (1981) Using animals to understand disease: A history. *Lab Anim (NY) 10*,2:30–37 [AB].

[68] Randall WC (1983a) Is medical research in jeopardy? *Physiol 26*,2:73–77 [AB]; Randall WC (1983b) APS testimony on S.657: "Improved standards for laboratory animals act." *Physiol 26*,5:271–272 [AB]; Randall WC (1983c) Past president's address: Crises in physiological research. *Physiol 26*,6(suppl):351–

356 [AB]; American Physiological Society (1983) APS statement on NIH *Guide* revisions. *Physiol 26*,5:273–274 [AB].

[69] Gill TJ III (1985) The rat in biomedical research. *Physiol 28*,1:9–17 [AB]; King FA, Yarbrough CJ (1985) Medical and behavioral benefits from primate research. *Physiol 28*,2:75–87 [AB].

[70] Miller NE (1985) The value of behavioral research on animals. *Am Psychol 40*,4:423–440 [AB].

[71] Tender costly care. (1978) *MD 22*,5:57–61 [AB].

[72] Benison S (1970) In defense of medical research. *Harv Med Alumni Bull 44*,3:16–23 [AB].

[73] Friess SL (1983) Critical issues facing animal scientists. *J Anim Sci 56*,1:217–221 [AB]; Gallup GG Jr, Suarez SD (1985) Alternatives to the use of animals in psychological research. *Am Psychol 40*,10:1104–1111 [AB]; Irving GW III (1985) Regulations and guidelines for animal care: Problems and future concerns. In: Moberg GP (ed) *Animal Stress.* Bethesda, MD: American Physiological Society, pp 281–296 [AB].

[74] Aronson LR, Cooper M (1976) Animal welfare and scientific research. [letter] *Science 194*:784–785; Hanlon CR (1982) Antivivisection in new clothes. *Bull Am Coll Surg 67*,6:1 [AB].

[75] Coile DC, Miller NE (1984) How radical animal activists try to mislead humane people. *Am Psychol 39*,6:700–701 [AB].

[76] American Society for Artificial Internal Organs (1968) ASAIO statement concerning the use of animals for research in the field of artificial organs. *Trans Am Soc Artif Organs 14*:454 [AB].

[77] American Association for the History of Medicine (1966) Resolution adopted by the American Association for the History of Medicine at its 39th annual meeting at Rochester, Minnesota, May 14, 1966. *Bull Hist Med 40*,3:277–278 [AB].

[78] Griffin RJ Jr (1981) Animals in pharmaceutical research. *Am Pharm NS21*,6:14–18 [AB].

[79] American Veterinary Medical Association Animal Welfare Committee (1985) Research animal statement proposed. *J Am Vet Med Assoc 187*,10:991–992 [AB].

[80] Fox MW (1984) Op cit [note 33].

[81] Indrieri RJ, Walshaw SO, Walshaw R (1983) Student surgery. [letter] *J Am Vet Med Assoc 183*,1:6,8.

[82] Greenwald GS (1985) ACDP survey on the use of animals in teaching physiology. *Physiol 28*,6:478–480 [AB].

[83] Rowan AN (1984) Op cit, p 261 [note 34].

[84] Ibid, p 177.

[85] Stark DM, Shopsis C (1983) Developing alternative assay systems for toxicity testing. In: Sechzer JA (ed) Op cit [note 6], pp 92–103 [AB].

[86] McMillan FD, Buyukmihci N (1983) Humane-oriented legislation. [letter] *J Am Vet Med Assoc 182*,1:5–6.

[87] Cohen BJ (1981) Animal rights and animal experimentation. In: Basson MD (ed) *Rights and responsibilities in modern medicine: Progress in clinical and biological research,* vol 50. New York: Alan R. Liss, pp 67–68, 85–92 [AB].

[88] Samuels WM (1984) Public affairs: Congressional amendment proposed by APS would punish lab intruders. *Physiol 27*,5:351 [AB].

[89] Held JR (1980) Animals in the lab. [letter] *Science 209*:214.
[90] Gallup GG Jr, Suarez SD (1985) Op cit [note 73].
[91] Cooper JAD (1983) Animals in biomedical and behavioral research. *J Med Educ 58*,1:71–72 [AB].
[92] Locke RF (1966) Legislation and research. *Ill Dent J 35*,6:331–332 [AB].
[93] Samuels W (1981) Op cit [note 3].
[94] Cohen BJ (1981) Op cit [note 87].
[95] Randall WC (1982b) Op cit [note 68].
[96] Cooper T, Galligan JE (1983) Leaving the island behind: An approach to the animal experimentation issue. *Physiol 26*,6(suppl):359–360 [AB].
[97] Samuels WM (1984) Public affairs: Reporters and researchers both to blame in telling the real story. *Physiol 27*,6:403–404 [AB].
[98] Anchel M (1976) Op cit [note 15].
[99] Anchel M (1977) Animals and ethics. [letter] *Science 195*:131.
[100] Neil DH (1981) The genesis of the Schroeder amendments. *Lab Anim* (NY) *10*,1:32–42 [AB].
[101] Kelly JA (1982) Disagreement with Gallistel: The need for increased scrutiny of animal research. [letter] *Am Psychol 37*,5:596–598.
[102] American Veterinary Medical Association (1983a) Hearings on animal welfare legislation. *J Am Vet Med Assoc 182*,5:466.
[103] American Veterinary Medical Association (1983b) Council endorses animal welfare amendment. *J Am Vet Med Assoc 183*,2:169–170.
[104] American Physiological Society (1983) If laws are needed, APS Council favors proposal by Sen. Dole. *Physiol 26*,3:113.
[105] Randall WC (1983b) Op cit [note 68].
[106] Samuels WM (1985) Public affairs: Congress establishes broader national policy for lab animal welfare. *Physiol 28*,6:477.
[107] WGBH Public Television (1975) Primate and the scientific ethos. Can some knowledge simply cost too much? *Hastings Ctr Rept 5*,1:6–8 [AB].
[108] Gallup GG Jr, Suarez SD (1980) On the use of animals in psychological research. *Psychol Rec 30*:211–218 [AB].
[109] Jacobs FS (1984) A perspective on animal rights and domestic animals. *J Am Vet Med Assoc 184*,11:1344–1345 [AB].
[110] The rights of animals: Morality or practicality? (1981) *MD 25*,10:98–103,107–108 [AB].
[111] Rowan AN (1984) Op cit [note 34].
[112] Hodos W (1983) Animal welfare considerations in neuroscience research. In: Sechzer JA (ed) *The role of animals in biomedical research.* Annals of the New York Academy of Sciences, vol 406. New York: The New York Academy of Sciences, pp 119–127 [AB]; Sechzer JA (1981) The use of animals for experimentation in the United States. *Psychopharmacol Bull 17*,2:88–91 [AB].
[113] Dubner R (1983) Pain research in animals. In: Sechzer JA (ed) *The role of animals in biomedical research.* Annals of the New York Academy of Sciences, vol 406. New York: The New York Academy of Sciences, pp 128–132 [AB]; Sechzer JA (1983) The ethical dilemma of some classical animal experiments. In: Sechzer JA (ed) *The role of animals in biomedical research.* Annals of the New York Academy of Sciences, vol 406. New York: The New York Academy of Sciences, pp 5–12 [AB].

[114] Silverman P (1978) Op cit [note 40]; Golley FB (1982) Editorial responsibilities in animal research. In: Dodds JW, Orlans FB (eds) *Scientific perspectives on animal welfare.* New York: Academic Press, pp 95–100 [AB]; Eyzaguirre CE (1982) Responsibilities of funding agencies for animal experimentation. In: Ibid, pp 69–76 [AB]; Iggo A (1979) Experimental study of pain in animals: Ethical aspects. In: Bonica JJ, Liebeskind JC, Albe-Fessard DG (eds) *Advances in pain research and therapy* vol 3. Proceedings of the 2nd World Congress on Pain, Montreal, Aug 27-Sept 1, 1978. New York: Raven Press, pp 773–778 [AB]; Rowsell HC (1977) The ethics of biomedical experimentation. In: NRC-NAS, ILAR, op cit [note 9], pp 267–281 [AB].
[115] McMillan FD, Buyukmihci N (1983) Op cit [note 86].
[116] American Veterinary Medical Association Animal Welfare Committee (1982) Veterinarian awareness of animal welfare issues seen as major priority by committee. *J Am Vet Med Assoc 180*,9:993–994.
[117] Wolff NC (1983) Op cit [note 58].
[118] American Veterinary Medical Association Animal Welfare Committee (1984) Animal welfare. *J Am Vet Med Assoc 184*,4:393–394. The AVMA's refusal to endorse the concept of animal rights was still drawing fire from veterinarians in 1987, when an article endorsing "a nonextreme position in the advocacy of animal rights" was published in *JAVMA* (Kronfeld DS, Parr CP [1987] Ecologic and symbiotic approaches to animal welfare, animal rights, and human responsibility. *J Am Vet Med Assoc 191*,6:660–664).
[119] American Veterinary Medical Association Council on Research (1963) Council report—Report of the AVMA panel on euthanasia. *J Am Vet Med Assoc 142*,2:162–170.
[120] American Veterinary Medical Association Council on Research (1986) 1986 report of the AVMA panel on euthanasia. *J Am Vet Med Assoc 188*,3:252–268. An edited version of the report was included in an appendix to the Interdisciplinary Principles and Guidelines of the New York Academy of Sciences' Ad Hoc Committee on Animal Research, reproduced in this volume in Appendix C.
[121] Sperling F (1966) Animal care in the laboratory: Who should regulate it? [letter] *Science 151*:776,778.
[122] Dennis C (1966) America's Littlewood crisis. The sentimental threat to animal research. *Surg 60*,4:827–839 [AB].
[123] Finch CO (1972) Trends in humane animal care legislation. *J Am Vet Med Assoc 161*,11:1533–1535 [AB].
[124] Schwindaman DF (1976) Op cit [note 15].
[125] Sechzer JA (1981) Op cit [note 112].
[126] De Paolo R (1981) Editor's note. *MD 25*,10:11 [AB].
[127] Harriton L (1981) Op cit [note 28].
[128] Holden C (1982a) New focus on replacing animals in the lab. *Science 215*:35–38 [AB].
[129] For example, Rosner F (1984) Animal experimentation. The Jewish view. *Arch Intern Med 144*,5:927–928 [AB]; Zola JC, Sechzer JA, Sieber JE, Griffin A (1984) Animal experimentation: Issues for the 1980s. *Sci Tech Human Values 9*,2:40–50 [AB]; Dobozy VS (1984) Comments on animal rights. [letter] *J Am Vet Med Assoc 185*,4:387.

130 Moss TH (1984) The modern politics of laboratory animal use. *Sci Tech Human Values* 9,2:51–56 [AB].
131 Holden C (1982) Scientists start fund to fight antivivisectionists. *Science 215*:640.
132 Hanlon CR (1982) Op cit [note 74].
133 Cohen BJ (1984) Public affairs: Pound animals for research and education. *Physiol 27*,4:190–191 [AB].
134 Ibid.
135 Myers D (1984) Animal rights, academic freedom, and progress in biomedical research. *Cancer Investig 2*,6:497–498 [AB].
136 For example, Trull FL, Kalikow NA (1984) Op cit [note 4]; Randall WC (1983a) Op cit [note 68]; Friess SL (1983) Op cit [note 73]; Irving GW III (1985) Op cit [note 73]; Kimmel HD (1982) Open letter to Dr. Edward Taub. [letter] *Pavlov J Biol Sci 17*,2:106; Campbell SL (1984) Animal research. *Science 224*:1043.
137 For example, Adams PM (1982) Investigator responsibilities in animal experimentation. In: Dodds JW, Orlans FB (eds) Op cit [note 114], pp 39–42 [AB]; Baker HJ, Lindsey JR, DaRif CA (1982) Responsibilities of institutions for the welfare of experimental animals. In: Ibid, pp 49–53 [AB]; Simmonds RC (1982) Summary of workshop on institutional responsibilities. In: Ibid, pp 63–65 [AB]; Halvorson HO (1982) The responsibility of scientists in determination of public policy for the use of animals in biomedical research. In: Ibid, pp 109–121 [AB].
138 Dodds WJ, Orlans FB (1982) Summary of recommendations on animal experimentation. In: Ibid, pp 123–126 [AB].
139 Cowley G et al (1988) Of pain and progress. *Newsweek 112*,26(Dec 26):50–57.

CHAPTER 6

Implications: The Challenge to the Scientific Community and the Role of Education

A Challenge to the Scientific Community

The success or failure of the new American animal research policy appears to rest on the effectiveness of the Institutional Animal Care and Use Committees (IACUCs). This is a formidable challenge to the scientific community: to investigators developing research proposals and to the IACUC members, most of whom are also researchers themselves. NIH, in deciding policy for the Public Health Service (PHS), continues in its reluctance to take on a policing role and shows confidence that self-regulation by scientists under a legislative mandate is the most realistic approach to take at the present time. By placing the responsibility on the institution and the IACUC for implementing and monitoring the new policy and for complying with the requirements of the Animal Welfare Act, the NIH has taken an approach similar to that of Canada and (until recently) Sweden, in its reliance on local ethical committees. Whether or not the IACUCs can indeed function as local ethical committees and effectively monitor research at their institutions depends on the adaptation of scientists to this drastic change in accountability.

The qualitative analysis of the scientific literature in the preceding chapter has allowed us to view the changing response of the scientific community to concerns about the use of animals for experimentation expressed by animal protection groups and the general public over the past 2 decades. During 1986, these responses and attitudes began to be tested, shaped, and reshaped as investigators and institutions approached their first year under the new PHS Policy and the amended Animal Welfare Act.[1] Attention has focused on the activities of the newly constituted IACUCs established at research institutions covered by the above mandates.

At first there was a great deal of resistance to IACUCs, even at many institutions accredited by AAALAC,[2] where animal care committees were already in existence, albeit not clearly defined. The IACUC's role in implementing the new policy and approving protocols was considered by many as an invasion of the scientists' privacy. To some extent, resistance was augmented by the fact that researchers had not previously been asked to address the humane treatment of the animal in protocols. This was now required, especially in cases involving more than minimal pain and other forms of distress to the animals, where restraint of animals was used, where multiple procedures were to be conducted on a single animal, or where death was anticipated as part of the experiment. There was also some alarm about the authority suddenly vested in the institution's veterinarian, who had never before had the power to make decisions involving an investigator's planned or ongoing experiments. These were some of the anxieties contributing to a generalized fear that the new system would act to restrict research.

In Chapter 5 we noted that ethical concerns have become a major theme and that this is a new way of approaching animal research for the scientific community. Examination of responses to the new policy confirms the lack of consensus we found concerning what ethical issues are important and how they can be addressed without compromising the scientific method and scientific validity. The recognition and alleviation of pain, fear, anxiety and other forms of distress are issues that illustrate the kinds of practical problems that are now surfacing.[3] The task of the IACUC in balancing animal welfare with scientific progress without compromising scientific validity is highly complex, and made more so by our limited understanding of animal pain, its recognition and alleviation. There is a need for the IACUC to assume an educational function in gathering information about species-specific behaviors of laboratory animals that might indicate that an animal is distressed. The work of the IACUCs has already been aided in this regard by the efforts of the American Veterinary Medical Association in sponsoring symposia such as the 1987 colloquium on animal pain referred to previously.[4]

In a related area, guidelines for methods of euthanasia continue to plague both investigators and those serving on IACUCs since the first year under the new policy. For instance, one recommendation of the 1986 Report of the AVMA Panel on Euthanasia, now included in the new PHS Policy, is that animals be lightly anesthetized prior to decapitation.[5] This specific recommendation was based on research published over a decade ago that suggested that animals could experience some pain and distress for as much as 13 to 14 seconds after decapitation.[6] When the 1986 AVMA report became part of PHS Policy, a major controversy erupted. Many scientists hotly contested the data and sought justification for an exemption.[7] Researchers worry about the effect of anesthesia on the validity of their results, because such drugs alter measurements of many variables, particularly those of neurotransmitters, immune factors, and enzymes.

Before the new PHS Policy took effect, the usual procedure was to decapitate an animal without anesthesia. As a result of the new rules, many investigators are now attempting to identify analgesics and anesthetics that do not interfere with the measurements under investigation, while at the same time requesting exemptions from using anesthesia for decapitation with scientific justification.

The IACUC was also perceived by many as a peer review group (ie, one that would evaluate the scientific merit of proposals), although the committee is not charged with this responsibility by PHS Policy or the Animal Welfare Act.[8] Yet, the more painful or distressful an experiment may be, the more important becomes its scientific justification. In evaluating the potential significance of proposed research, we are faced with an epistemological question: How do we know the ultimate value of any research, especially before it has been carried out? The following statement, proposed as a moral yardstick for scientists, assumes what is really highly problematic:

> Animals should not be used in painful experiments when substantial benefits are not expected to result. Even when the objective is important, there is a presumption against the use of animals in painful and dangerous experiments that are expected to yield tenuous results of doubtful value. Animals but not human beings may be used in painful and dangerous experiments that are to yield vital benefits for human beings (or other animals).[9]

This statement is based on a presumption that it is possible to tell in advance whether the outcome of the experiment will be significant. Has our knowledge of science become sophisticated enough to make such predictions? As long as animals are used for research, the scientific community will continue to be faced with the ethical dilemma of deciding to conduct a possibly painful experiment whose significance cannot be predetermined or to choose not to conduct an experiment because it may be painful and so possibly miss a significant finding.[10]

At a workshop held at the Hastings Center in New York, scientists argued against this additional responsibility for the IACUC because they felt that the committee could not conduct an appropriate peer review.[11] Consequently, workshop participants felt that an investigator would have good cause to appeal a negative decision by the IACUC on his/her protocol if the decision were not based solely on the humane aspects of animal treatment. There was, however, general agreement on the principle that the greater the potential pain or distress an experiment has, the greater the tendency or need to consider the scientific merit of the experiment. Nevertheless, many IACUC members caution the PHS against making this a responsibility of the IACUC since it would create more tension between the researchers, the IACUC, and the PHS and would raise serious questions about the capability of the committees to make such judgements.

As the IACUCs enter their fourth year of activity many of the tensions and concerns have been considerably reduced. However, it is still somewhat premature to say whether the new policy has interfered with the spontaneity of research or has impaired the ability of scientists to conduct a specific type of experiment or to use a particular species. Some scientists feel that it has, whereas others feel that as a result of IACUC actitivies, the treatment of animals—and therefore the quality of the scientific data—has improved.

There are several factors that will help to determine the effectiveness of the IACUC and PHS Policy in assuring the humane treatment of laboratory animals. The first is how IACUCs consider and manage future problems in animal use. IACUCs are and will be faced with situations reflecting new developments in science and technology that require the use of animals, some in entirely new ways. IACUCs are beginning to make decisions about the preparation and use of transgenic animals and will be faced with ethical questions regarding the preservation of species integrity. Another area of concern is the use of procedures involving the transplantation of animal organs to humans for lifesaving purposes. These procedures are currently being carried out on a small scale, and if successful they will be conducted in increasing numbers. Such procedures are still innovative and require Institutional Review Board approval as well as PHS approval. At present the role of the IACUC regarding such transplants falls into a grey area, determined by the boundaries set for the IACUC by the institution and the administration's view of the procedures. Some institutions have extended PHS Policy to cover all uses of animals including teaching, regardless of the source of funding. At such institutions, if the transplant procedure is viewed as an experimental or innovative procedure, the IACUC would most likely be asked to review and approve the proposed transplant, especially in cases where the donor animal belongs to the institution.

A second factor is the persistence of activities of animal protection groups, especially animal rights organizations. There have been continued pressures from such groups to abolish the use of specific procedures on animals as well as to halt the use of certain species. Some of these efforts have been very effective in stopping particular research projects targeted by the activists, as noted in Chapter 4 (see endnote 6). Animal protection groups have also begun to use legal channels to try to gain access to the minutes and records of the IACUCs to determine whether the committees are fulfilling their responsibilities to assure the humane care and treatment of the animals in accordance with PHS Policy and the Animal Welfare Act.

Another issue affecting research is the use of random-source, or pound, animals (as opposed to animals bred for research). According to a recent study by the National Research Council, concern over the use of pound animals (dogs and cats) for experimentation is a major issue in the controversy. The use of pound animals has been challenged by animal welfare advocates who feel that former pets experience more stress in the research laboratory than purpose-bred animals do, and that for this reason they should not be used for research. Animal protectionists also argue that the results of experiments utilizing pound animals are questionable because of the varied genetic backgrounds of these animals. Whether pound animals should be used at all, used only under anesthesia in experiments from which they will not recover, or used in any experiment for which they are considered suitable are questions that remain to be resolved. At present, 49 states permit pound animals to be used for experimentation, 11 states prohibit the release of animals by pounds for experimentation but allow researchers to obtain pound animals from licensed dealers in other states, and one state—Massachusetts—forbids the use of pound animals regardless of their state of origin.[12]

Concomitant with these pressures is the issue of who should be permitted to do what to an animal, a question that is already under discussion. Some feel that the United States should develop a system comparable to the three-tier system recently introduced in the United Kingdom, in which three types of licenses are issued permitting animal experiments to be performed at different levels of complexity based on the individual's qualifications.[13]

It may be necessary in the future to change the charge to the IACUC or to make major changes in PHS Policy. Nevertheless, considerable progress has been made, notably in institutions that have never before had any type of animal care committee. It will take many more years to decide with certainty the success of the IACUC and of the new American policy. At any rate, in the event that PHS Policy fails, it would surely open the door for more highly restrictive legislation, possibly following the lead recently taken by the United Kingdom (described in Chapter 7).

The effects of the newly enacted legislation are also being felt outside the United States. In the most recent revision of PHS Policy, the following statement appears: "Institutions in foreign countries receiving PHS support for activities involving animals shall comply with this Policy, or provide evidence to the PHS that acceptable standards for the humane care and use of animals in PHS-conducted or supported activities will be met."[14] Just what such "acceptable standards" are has not yet been made clear. The new PHS Policy may in fact pose a serious problem for the funding of animal experiments in foreign countries whose laws, regulations, or guidelines, particularly with regard to the use of primates, are not comparable with the new American policy. The PHS has long supported research outside the United States, a practice that has tended to increase over the years. Precise data about the extent of such support have been difficult to obtain. However, between 1986, when the new PHS Policy was implemented, and late 1988, 139 assurances for animal research programs in foreign countries were approved by NIH's Office for Protection from Research Risks (OPRR). This is in comparison with more than 1,000 approved assurances on file from American institutions.[15] Whether this represents a decrease in foreign applications because of the new regulations remains to be determined. It would seem also that third world countries, just beginning to develop or expand their research enterprises (many of which involve testing), will have the greatest difficulty in providing appropriate evidence for acceptable practices, not only because of economic conditions, but also because in many of the countries laws or other policies for animal experimentation have not yet been established.

Another potential effect of the recent legislation is the possibility that editors of American scientific journals will refuse to publish research reports unless the research is carried out in compliance with the new policy. Some journals (eg, *Pain* and the *Journal of Endocrinology*) have already made this stipulation in their editorial guidelines. This could affect the small number of investigators at institutions that are not directly covered by the federal legislation: those that do not receive PHS support or buy animals in interstate commerce, or—as is the case in some industrial organizations—use only rats and mice. Such an editorial policy would also affect investigators in countries with less restrictive regulations than ours. Conversely, our new policy could result in more publications by American scientists in journals of other countries whose laws were previously much more restrictive than ours. Some British and Scandinavian journals, for instance, publish results of scientific research only if the conditions under which the experiments were performed meet those mandated by the country in which the journal is published. In the past, many of these editors have rejected American manuscripts, especially those reports of research in which neuromuscular-blocking (paralytic) agents were used without anesthesia.

The Role of Education

Now that we have two major pieces of legislation in the United States regulating the use of animals for experimentation, we need to decide what the next step should be to further humane ethics and animal welfare. The issue of the use of animals for research, testing, and educational purposes will surely be with us for a very long time, and efforts to restrict and/or abolish the use of certain species or particular types of experiments will increase, not just in the United States, but on an international scale.

Can we balance animal protection with scientific progress? Because of the size of our country and the very large number of investigators in the American research community, the United States is confronted with a unique problem in assuring continuing compliance of investigators with the laws and PHS Policy. One of the most important ways to assure the continuity of animal use for the benefit of both humans and animals is to develop comprehensive educational programs for students, young scientists, and others who will be involved in any aspect of animal care and use.

Both of the 1985 laws contain new requirements for training. As stated in the Health Research Extension Act, "scientists, animal technicians, and other personnel involved with animal care, treatment, and use by the applicant [organization] must have available to them instruction or training in the humane practice of animal maintenance and experimentation, and the concept, availability, and the use of research or testing methods that limit the use of animals or limit animal distress."[16] Yet, at a recent meeting of an NIH workshop in Virginia, it was pointed out that, while some training programs are available for technicians and auxiliary personnel, there are almost none at present for the scientists.[17] Several scientific societies have held talks, symposia, or workshops on current issues (including ethical concerns) and research techniques involving animals, but the need for basic and comprehensive educational programs within the curriculum for young scientists persists.

To be effective and generate enduring results, an educational program must cross the broad spectrum of scientific disciplines. Students and beginning scientists must learn about the most humane methods and techniques available for conducting experiments when animal use is required, so that they will *think* when they design their experiments and ensure that pain and other forms of distress are prevented or reduced to the necessary minimum. An understanding of the physiology of the animal and its species-specific behaviors will aid in the recognition of signs that could indicate when an animal is in distress. Students should also learn about available alternatives for replacing animals in some kinds of experiments, and they should be taught the appropriate statistical techniques necessary to calculate the minimum number of animals required to determine the significance of their results.

Every educational program should also include material on philosophy and ethics. Students need to develop some understanding of the important moral and ethical problems posed by animal experimentation and the responses that concerned people have made. It is also important that students learn how to use some basic philosophical skills so that they may distinguish between good arguments and bad ones, and acquire the ability to comprehend and analyze the different philosophical and ethical positions about animal research.[18] As part of any educational program, students should become familiar with the history of animal use and contributions made to human and animal life and health through experiments on animals. This material should be included in the curricula of undergraduate institutions and of graduate and professional schools involved in the training of scientists in all disciplines that use live animals for research, testing, or educational purposes.

Such an educational approach is particularly important in view of the heightened concern about animal use by students in both undergraduate and graduate programs. Many students are questioning the propriety of animal use in education or are refusing to take part in studies or exercises requiring demonstrations on live animals. Many states are currently considering laws giving students permission to absent themselves from such experiments without losing course credit.[19] Consideration needs to be given to such problems in order to find practical solutions that allow students to develop their moral sensibilities without forfeiting valuable educational information.

The development of a comprehensive educational program was one of the primary objectives of the Ad Hoc Committee on Animal Research of the New York Academy of Sciences, whose work was mentioned briefly in the Introduction. In consultation with an advisory panel of representatives of 21 scientific societies and organizations, the group worked for more than 2 years to produce a set of "Interdisciplinary Principles and Guidelines for the Use of Animals in Research, Testing, and Education." It was the intent of the committee and the advisory panel to provide a standard for animal use that scientists and local review committees could use as a reference tool, and upon which institutions could base comprehensive educational programs. The document, which addresses major concerns about the use of animals in experimental procedures, is reproduced in this book in Appendix C.

When an educational program such as the one envisioned here is introduced as a serious part of the curriculum, it should inculcate in the next generation of scientists and technicians a sense of what Caplan has termed "moral stewardship."[20] In addition, such a program would help stimulate a constructive dialogue with moderate animal welfare groups, which continue to challenge the scientific community. Finally, the development of an extensive educational program throughout undergraduate and graduate schools in this country would go a long way toward encouraging the growth of a system that can adjust well to problems—ethical and otherwise—engendered by future developments in science and technology.

Notes

(The notation *[AB]* after a reference indicates that it is included in the Annotated Bibliography following Chapter 7.)

1 Prior to the passage of legislation, representatives of scientific societies and organizations had testified in Congress, addressing various points in the proposed laws. Many societies revised or reemphasized their guidelines during this time, and held symposia to discuss and debate proposed regulations. This helped to prevent hasty legislation and helped Congress take a balanced approach to the pending bills. During 1985 and 1986, the United States Congressional Office of Technology Assessment organized an advisory panel to address the issue "Alternatives to Animal Use in Research, Testing, and Education." Meanwhile, research on alternatives increased through funding by government and private agencies. See U.S. Congress, Office of Technology Assessment (1986) *Alternatives to animal use in research, testing, and education.* Washington, DC: U.S.

Government Printing Office, OTA-BA-273; Public and private funding toward the development of alternatives. In: Ibid, pp 259–272.

[2] See Chapter 1 for a description of the accreditation program of the American Association for Accreditation of Laboratory Animal Care (AAALAC).

[3] For working definitions and discussions of these terms see Kitchen H, Aronson AL, Bittle JL, McPherson CW, Morton DB, Pakes SP, Rollin B, Rowan AN, Sechzer JA, Vanderlip JE, Will JA, Clark AS, Gloyd JS (1987) Panel report on the colloquium on recognition and alleviation of animal pain and distress. *JAVMA 191*,10:1186–1191.

[4] AVMA colloquium on recognition and alleviation of animal pain and distress. (1987) *J Am Vet Med Assoc 191*,10:1186–1298. This problem is currently being considered by a National Research Council/Institute of Laboratory Animal Resources Committee on Pain and Distress in Laboratory Animals.

[5] American Veterinary Medical Association Council on Research (1986) 1986 report of the AVMA panel on euthanasia. *J Am Vet Med Assoc 188*,3:252–268.

[6] Mikeska JA, Klemm WR (1975) EEG evaluation of humaneness of asphyxia and decapitation euthanasia of the laboratory rat. *Lab Anim Sci 25*,2:175–179 [AB].

[7] For example, Allred JB, Berntson GG (1986) Is euthanasia of rats by decapitation inhumane? *J Nutr 116*:1859–1861. H.C. Rowsell has brought to our attention a letter from the chairman of the 1986 AVMA Panel on Euthanasia, protesting that the panel's report was never intended as a "statute to be applied ... in an entirely rigid way," but rather the report was written as a guide, in cognizance of the fact that sufficient information is not yet available for a definitive statement on the subject (Smith AW [1988] Euthanasia. *J Am Vet Med Assoc 193*,2:161–162). Nevertheless, PHS Policy does require that "methods of euthanasia used will be consistent with the recommendations of the American Veterinary Medical Association (AVMA) Panel on Euthanasia, unless a deviation is justified in writing by the investigator." See Office for Protection from Research Risks (Sept 1986) *Public Health Service Policy on humane care and use of laboratory animals.* Bethesda, MD: National Institutes of Health, Sec IV.C.1.g. In the letter cited above, Smith urges review committees to "take the responsibility of rendering appropriate decisions by balancing the research need against the perceived possibility that a decapitated animal could suffer 13 to 14 seconds of pain." The New York Academy of Sciences' Ad Hoc Committee on Animal Research addressed this question in its *Interdisciplinary principles and guidelines for the use of animals in research, testing, and education* (1988). The committee recommended that "when, for scientific reasons, the procedure cannot be conducted in accordance with AVMA guidelines, it should be regarded as part of the experimental procedure with specific approval of the full IACUC" (p 9).

[8] In New York State, however, the committee is specifically charged to review the scientific justification of the research. See New York State Department of Health, Center for Laboratories and Research, Chap II, Administrative Rules and Regulations, Subchapter D, Part 55 (1.4 b), effective July 1, 1983.

[9] Hoff C (1980) Immoral and moral uses of animals. *N Engl J Med 302*,2:117–118 [AB].

[10] Sechzer JA (1983) The ethical dilemma of some classical animal experiments. In: Sechzer JA (ed) *The role of animals in biomedical research.* Annals of the New York Academy of Sciences, vol 406. New York: The New York Academy of Sciences, p 11 [AB].

[11] This workshop, "Ethics and Animal Welfare," was organized by Arthur L. Caplan (The Hastings Center, Hastings-On-Hudson, New York) and held on April 8, 1987, to evaluate the adequacy of IACUCs for both protecting the welfare of animal subjects and allowing the conduct of useful research and testing with animals.

[12] National Research Council (1988) *Use of laboratory animals in biomedical and behavioral research.* Report of the committee on the use of laboratory animals in biomedical and behavioral research. Washington, DC: National Academy Press, pp 64–67. The committee unanimously recommended that pound animals be made available for acute experiments in which they will remain anesthetized until killed. The majority of the committee supported the use of pound animals in all experiments, but a minority opposed their use for chronic (survival) experiments. One dissenting committee member did not sign the report.

[13] This issue, as well as a discussion of the problems involving transgenic animal procedures, was raised by Dr. Gary Ellis, senior analyst at the Office of Technology Assessment, during a meeting of the Hastings Center Committee on Ethics of Animal Experimentation held Nov 1988. The three-tier system of licensing used in the United Kingdom is described in Chapter 7.

[14] OPRR (1986) Op cit [note 7], p 1.

[15] Personal communication from John Miller (Oct 1988), Office for Protection from Research Risks of the National Institutes of Health.

[16] The Public Health Service Act. Title 42 US Code, Chap 201 *et seq*, Public Law 78–184 (1944) amended by the Health Research Extension Act of 1985, Chapter 289d) sec 495a) Public Law 99–158: NIH Reauthorization (1985). See Appendix B.

[17] Workshop on the Care and Use of Laboratory Animals, Dec 1987, Charlottesville, VA.

[18] We thank Professor Lilly-Marlene Russow of Purdue University for sharing her thoughts and information on philosophical approaches to this problem.

[19] *Update* (various issues), a publication of the National Foundation for Biomedical Research. One widely publicized case involved a Los Angeles-area high school student who refused to dissect a frog in biology class. Her suit against the school district was dismissed, however, when the school agreed to let her identify frog anatomy from photographs of a dissected frog that had died of natural causes. (Student who sued school isn't required to cut frog, *New York Times*, Aug 8, 1988). See also Sieber JE (1986) Students' and scientists' attitudes on animal research. *Amer Biol Tchr 48*,2:85–91. This article presents the results of a survey that showed that undergraduate students felt a need for an ethical analysis in courses where animal dissection would be carried out, as well as a specific justification for the procedure. Students also felt that they should be given a choice about performing a dissection on an animal.

[20] Caplan AL (1984) Moral stewardship and the responsibility of scientists using animals in research. Presented at the 92nd Annual Convention of the American Psychological Association.

Chapter 7

The International Context: Regulation Around the World

Legislation in Europe and Asia

Denmark (1953)

Denmark passed its first law containing regulations for animal research in 1953. It stated, in general, that animal experimentation was prohibited unless specifically authorized (as was the case in other Scandinavian countries). All laboratories were required to be licensed. Painful experiments could be performed only under anesthesia unless the nature of the experiment precluded this. Special authorization was necessary for such procedures, and in cases where it was expected that pain would persist, it was required that the animals be killed while still under anesthesia. If an experimental procedure required that the animal recover, special authorization was required. For certain types of experiments, individual animals could be the subject of only one experiment. Detailed records of all experiments were required, and inspections were carried out by the Licensing Authority. There was also comprehensive legislation covering the transport of animals.[1]

The law was rewritten in 1977, principally by Professor Heed Adler of the Ministry of Justice (now deceased) with Professor Aawe Thordal-Christensen of the Small Animal Clinic in Copenhagen.[2] The 1977 law, under the Ministry of Justice, licenses individual investigators for experimental procedures that are expected to cause the animal more than momentary or minor pain. Licensing is not required for procedures causing only momentary or minor pain. If alternative methods are available, such as tissue or organ culture, then the use of animals for the same purpose is forbidden. The law covers only vertebrates and only animals of the lowest possible "rank" may be used.[3] Animals may be used for educational purposes, but only for training in experimental techniques. Since January 1, 1986, the law requires that dogs, cats, and rabbits intended for experimental use be purpose-bred. Anesthesia is required for all potentially painful procedures, and the animal is not allowed to recover from anesthesia unless the pain will not endure or recovery is essential for the scientific purpose of the experiment. The method for euthanasia for dogs, cats, and primates must be stated in advance. Records must be kept of the species and number of animals used, the procedures carried out, and the fate of the animals.[4] Since there are only approximately 300 licensed scientists in Denmark, implementation appears to be efficient and effective. It is expected that an additional requirement will be made for ethical review committees.

Poland (1959)

In Poland, animal experimentation is regulated under the Decree of the Minister of Higher Education of 1959 (which amended the Constitutional Act of 1928). The regulation provides that research institutions be exclusively licensed to perform experiments with animals; the institutions are authorized to grant licenses to individual investigators with university degrees. It would seem unlikely that researchers could circumvent this system of authorization, since prior approval and progress reports are required for all experimental projects. As of 1984, there were no additional statutory controls such as an inspection system like that in the United Kingdom or a regional committee system as in Sweden. Emphasis appears to be on the professional and ethical education of investigators and on use of the ICLAS (International Council for Laboratory Animal Science) recommendations, which have been translated into Polish and apparently widely disseminated. Although there is no required registration of the species and number of animals used, there is annual registration on a voluntary basis. According to a recent assessment, about two-thirds of the research centers report these data voluntarily.[5]

France (1963)

French legislation passed in 1963 pertained primarily to general animal welfare, but various articles and decrees have been issued to include animal experimentation. These decrees are under the Penal Code and are administered by the Ministry of Agriculture and other ministries that act in an advisory capacity. Experiments for scientific purposes are not to be undertaken unless there is reason to hope that they will lead to an "hitherto unconfirmed scientific conclusion" or throw light on unsolved problems. Experiments are individually authorized, and there are provisions governing food and housing. The law also states that care is necessary to make sure the animals are not subject to unnecessary suffering. Painful operations are to be carried out under anesthesia or an equivalent analgesic procedure. If the animal will not survive the procedure, it is to be killed while still under anesthesia. Unless justification is given, experiments without anesthesia may be performed only once on an animal. France is now working toward a consensus for possible new legislation.[6]

Luxembourg (1965)

The Animal Welfare Act of 1965 provides for protection of animals in Luxembourg. While most of the law is for general protection of animals, it nevertheless contains provisions relating to animal experimentation. Some of these cover proper housing for animals, including requirements for adequate space, ventilation, lighting, and protection from the elements. The transportation of animals in vehicles or containers is regulated. The act specifically states that it is an offense to carry out experiments on live animals that are of such a nature as to cause them prolonged suffering.[7] In 1983 the Law for the Protection and Welfare of Animals, under the authority of the Ministry of Agriculture, was passed to supercede the 1965 act. Licenses are issued on approval by veterinarians at the Ministry of Health, and inspections of facilities are conducted by veterinarians of the Ministry of Agriculture. Additional provisions are the same as those contained in Swiss law pertaining to animal experimentation.[8]

Finland (1971)

Laboratory animal activities in Finland are guided by the Finnish Animal Welfare Act of 1971 and its amendments. Animals may be used for scientific purposes with permission of the Veterinary Department of the Ministry of Agriculture and Forestry. Animal experiments may be carried out only in authorized institutions and under the supervision of the director of the institution. The ministry must be notified in advance of the details and purpose of the study as well as the species to be used. Painful experiments may be conducted only with general or local anesthesia "unless the experiment requires otherwise." Painful experiments may be undertaken only by medical and veterinary surgeons or by others authorized by the ministry. Animals may also be used for educational purposes when other methods "cannot be substituted." Records must be kept of the species, number of animals, and the purpose of the research. Inspections are made by the ministry, especially when painful procedures are used.

In January 1986 a new animal protection law was enacted with a statute covering animal experimentation. The main changes include (1) a requirement for an Establishment License to practice animal experimentation, (2) preparation of an experimental plan and application by the investigator for a license permitting the experiment, and (3) the establishment of local animal experiment committees and a classification of experiments. For all vertebrates, a detailed plan of the experiment must be submitted to the committee with an evaluation of any pain and suffering involved and methods of pain relief. Justification is required if pain will not be alleviated. Experiments are divided into two classes, to be determined by the committee. The first covers experiments such as LD50 tests, studies in which the animal's freedom is restricted, and procedures that may cause more than minimal pain or suffering, or serious illness. The second class includes procedures that may cause slight or short duration of pain, suffering, or distress. The committee can approve class 2 studies, but decisions about class 1 studies are made by the county government. Although the new system tightens controls, it does not "presuppose the representation of the Society for Prevention of Cruelty to Animals in the committees."[9]

Federal Republic of Germany (1972)

As a Federal Republic, Germany passed its first law concerning animals, the Animal Protection Act, in 1972. The Ministry of Food, Agriculture, and Forestry served as the administrative authority. Under this law, licenses were issued only to the heads of institutions. However, although the law is national, it is administered individually by the different states. Each state can use sanctions that vary from stopping an experiment and removing the animals, to revoking a permit, or imposing fines and a maximum of 2 years' imprisonment. It is not permitted to cause pain, injury, or suffering to an animal without justification. If invertebrates can be used to achieve the same purpose, vertebrates are not permitted; similarly, when cold-blooded animals can be used, warm-blooded animals are not permitted. Experiments are limited to the minimum number of animals necessary. Unlike regulations in the United Kingdom and Denmark, the German law allows animals to be used for educational purposes at the high school level without obtaining a permit. Such experiments, however, must be reported to the appropriate authorities prior to being carried out. Amendments were passed in 1982 and 1984.[10]

A new law, passed in 1986 and implemented in 1987, is much more restrictive. It begins, "It is the object of this law to protect, out of man's responsibility for the animal as his fellow creature, its life and well-being." Animal experiments are defined for purposes of the law as "operations or treatments applied to animals for experimental purposes, which may be associated with pain, suffering, or injury to the animal." In planning an experiment, the investigator should determine whether the goal of the experiment could be carried out with alternative methods or procedures. Vertebrates may be used for animal experiments only if they have been purpose-bred; if pain, suffering, or injury is anticipated, the procedure(s) must be ethically justified in view of the purpose of the study. Complete or partial amputation of parts of the body or the complete or partial removal or destruction of organs or tissues of a vertebrate is unlawful unless such procedure is necessary for the purpose of transplantation, developing cultures, or studying isolated organs, tissues, or cells. It is unlawful to use animals to develop or test weapons, ammunition, or similar equipment. Requests for approval of animal experiments must be written in detail and include the scientific justification. The law requires that a commission be set up in each state to approve proposed animal experiments. Such commissions may operate as ethical review boards. Research facilities must appoint one or more agents (an animal welfare officer) charged with animal protection and report the appointment to the competent authority. In experiments where a lethal dose is to be determined, the animal must be killed painlessly as soon as it is evident that it will die in order to avoid prolonged suffering.[11]

Iceland (1973)

Iceland passed its first law concerning animal research in 1973. It is administered by the Chief Veterinary Officer in the Ministry of Culture and Education. Experiments may be carried out only in approved institutions and only by persons with a special permit. Permits may be issued to veterinarians, physicians, or persons working in institutions with personnel having the required degrees. Appropriate anesthesia must be administered, with euthanasia at the end of the experiment, and records of all procedures must be kept. Legislation is currently under revision.[12]

Japan (1973)

Japan's 1973 law, implemented in 1974, includes standards relating to animal protection. The law, Concerning the Protection and Control of Animals, was enacted "to prevent cruelty to animals; to provide for appropriate treatment, taking natural habits into account; to engender a feeling of love for animals among people, thereby contributing to the development of respect for life and sentiments of friendship and peace; and to protect humans from any hazards to themselves or their property that could result from possession by others of domestic, farm, laboratory animals, or wild animals in captivity." The law establishes a fine of up to ¥30,000 (about $223 at the time of writing) for violations of the law or of standards implementing it. The law covers mammals and birds, but is intended to apply to other species. The regulations outline responsibilities applicable to laboratory animals, mainly concerning anesthetics and euthanasia. Owners of animals are responsible for their maintenance, health, and safety. In Japan, procedures may be carried out with animals for research, for the manufacture of antibiotics, for education, and for other scientific purposes. When such procedures are conducted, the law states that animals should suffer the minimum pain possible within limits of the purpose of the study. If an animal will not recover from a scientific procedure, the person conducting the research must immediately dispose of the animal using the method causing the least amount of pain.

A new law passed in 1986 pertains to the care and management of farm animals. Its principles are the same as those of the 1973 law but include specific regulations for the maintenance of health and safety of farm animals, transportation of farm animals, prevention of injury to farm animals, and conservation of the living environment.[13]

Austria (1974)

A federal law on animal experimentation was passed in Austria in 1974. The law is administered by different ministries, depending on the type of research facility in which the experiments are conducted. Permits for specific types of experimental procedures with animals are granted to the heads of institutions. Anesthesia is required for painful experiments unless it "frustrates" the purpose of the study, and surgical procedures on vertebrates may be carried out only by suitably qualified scientists. Inspections of the facilities are made by the relevant authority. Records must be kept for 2 years and must include the purpose of the experiment, the number of animals, and their origin.[14]

Norway (1974)

In Norway, the Welfare of Animals Act, passed in 1974 but not enacted until 1977, was the country's first law pertaining solely to the use of laboratory animals. As early as 1935, Norway had passed the Protection of Animals Act, which gave animals very extensive general protection but did not specifically concern the use of animals for experimental purposes. Another act, passed in 1972, extended this protection and included animals transported by air. The 1974 act contained some regulations concerning experimentation and the use of laboratory animals that were in line with other Scandinavian legislation.[15] Scandinavian countries, in general, have shown much concern about the infliction of pain on animals; the recent law reflects this concern (as do the laws of Denmark, Finland, and Sweden). The act, administered by an Experimental Animals Board (EAB) appointed by the Ministry of Agriculture, covers all vertebrates and crustacea. The EAB issues and enforces regulations covering the use of animals for experimentation. Regulations provide that animals shall be treated well, with due regard given to their natural instincts and needs so that they will not suffer unnecessarily. Licenses are granted to approved institutions with qualified directors who assume responsibility for all personnel, although under special circumstances an individual license may be issued.[16]

In 1985 the law was amended so that no experiment on a live animal may be carried out without the written consent of the director of the institute, with final approval obtained from the EAB. If alternatives are available, licenses will be refused. When animals are used, there are specific requirements for anesthesia, analgesia, and euthanasia. Special permission is needed for painful experiments without anesthesia, and for all experiments on dogs, cats, and nonhuman primates. Requirements include an annual report to the EAB by the director of the institution providing information on the number of animals used, the species, the type of experiments carried out, and the disposition of the animal at the end of the experiment.[17]

Belgium (1975)

A Protection of Animals Act, passed in 1910, dealt with the general protection of animals, including some aspects of animal experimentation. The 1910 law regulated space allowance and ventilation during transport and methods for tethering animals while in transit.[18] Although there was no legislation controlling surgical operations on unanesthetized animals, animal experimentation was regulated, and all laboratories had to be licensed by the state. It is unclear how the 1910 law was implemented and enforced or the amount of experimentation carried out. In 1975, however, the Animal Protection Act was passed specifically for the use of animals in experimentation. It is administered by the Ministry of Justice and Ministry of Agriculture. Directors of laboratories are licensed by the Office of Veterinary Medical Inspection, and inspections are conducted by state veterinary officers. Anesthesia is required for all painful procedures, unless it will interfere with the purpose of the experiment. A 1981 decree has added the requirement that students in veterinary, medical, and biology programs be trained in alternative methods. New legislation is under consideration.[19]

The Netherlands (1977)

The Netherlands passed the Law for Animal Experiments in 1977. As in most European and American countries, a general law protecting animals from cruelty had been in existence since the late 19th century. This legislation was extended to cover some aspects of research when the Animal Protection Act of 1961 was passed. Applicable to animal experimentation was the statute regulating trade in or board of dogs and cats, or training, selling, or exhibiting animals other than domestic species and wild animals native to The Netherlands.[20] However, legislation pertaining solely to animal experimentation did not exist until the 1977 law was passed. The Veterinary Chief Inspectorate of the Ministry of Public Health has the responsibility for administering the law: registering research facilities, making periodic inspections, issuing regulations and guidelines, and regulating sources of laboratory animals. The law has established a most comprehensive system for regulating animal experimentation and became the model for the Council of Europe's Draft Convention on protection of laboratory animals. The responsibility for implementing the law is given to the heads of the research institutions to whom licenses are issued. The head of the institute is responsible for appointing experts to ensure that all personnel involved in animal research are qualified to perform the assigned tasks and that the welfare of the animals is supervised by a qualified veterinary surgeon or equivalent professional. Scientific justification is required for all animal experiments on vertebrates that are injurious or are likely to cause significant pain or distress. Painful experiments on vertebrates are not permitted when alternative methods are available, and anesthetics are required for more than minimal pain, except when their use would frustrate (sic) the object of the experiment. If severe and prolonged pain is likely to result, the animal must be humanely killed before it recovers from anesthesia. The Netherlands law, like the British law, has instituted a rigorous system for assessing laboratory animal use. This system produces accurate data about the number of animals used as well as the nature of the research.[21]

Since January 1986, licensed institutions are required to have an institutional ethics committee composed of persons representing different disciplines, including ethics, to oversee all experiments. Animals suffering pain must be euthanized once the experiment is completed. Alternatives must be used where available, and "lower" vertebrates substituted for "higher vertebrates." This pertains especially to primates, cats, dogs, and horses.[22]

Sweden (law amended 1979, 1982)

In Sweden, The Protection of Animals Act of 1944 was amended in 1979 and again in 1982. The regulations included requirements for individual licenses, the use of purpose-bred animals, and pain relief whenever anesthesia is not used. "Higher" and "lower" vertebrates are distinguished, and warm-blooded animals may not be used in teaching where other methods are available. Detailed recordkeeping is required, especially regarding the number of animals used. A system of classifying the "severity of experiments" was developed, based on a distinction between physiological pain and psychological distress. A regional ethical system of reviewing and approving research proposals was developed by the Laboratory Animals Board of the National Medical Research Council and implemented with the expertise of Professors Karl Öbrink, at the University of Uppsala's medical school, and Lars Wass, a representative of the National Board of Universities. At each of Sweden's six regional research centers, a local ethical committee was established for the purpose of reviewing and approving experimental protocols. One proposed advantage of such a system was that approval of an experiment could be achieved within a few days and thus not affect the spontaneity of research. This was in contrast to procedures for obtaining approvals in other countries, which often takes weeks or even months.[23] By 1986, it became apparent that the system was not a success. The regional ethical system has now been abolished, and new legislation for animal research is being developed.[24]

Farm animals, however, are the beneficiaries of a new animal welfare law enacted in July 1988, instituting "what is probably the world's strictest program of rights for farm animals," according to *Lab Animal*. Cattle, pigs, and chickens are freed from factory farm methods. Cattle have grazing rights, pigs cannot be tethered and must be "granted" separate bedding and feeding places. Pigs and cows must have access to straw and litter and chickens are to be let out of cramped cages. Drugs and hormones can be used only to treat disease. The law will be implemented over several years.[25]

Switzerland (1981)

The protection of animals is addressed in the Swiss constitution. By 1958 Switzerland had, in addition to the Swiss Penal Code (which gave general protection to animals), the Federal Road Traffic Act of 1958. The Traffic Act established rules for the transportation of animals.[26] Regulations directly addressing animal experimentation were passed in the federal law of 1978, Regarding the Protection of Animals (as amended by the Ordinance of 1981, Regarding the Protection of Animals). Because of pressures by an active and sometimes violent antivivisection movement, the Swiss academies of science developed additional guidelines in 1983; these have been adopted by the government, industry, and academia. Under these guidelines, a permanent committee was established to review experiments using animals, and special requirements were instituted for experiments that involve pain (only vertebrates are covered). The regulations are administered by a competent authority of each canton or state by a committee of specialists. Authorization is required for each experiment or series of related experiments, and all experiments must be carried out under the direction of a trained scientist. A special permit is required for experiments involving pain, distress, or disturbance of the general condition of the animal. Anesthesia is required for all potentially painful procedures, but exemptions can be obtained where justified. Where painful experiments are to be carried out, suffering is not to exceed that which is absolutely necessary. Wherever possible, "lower" species are preferred to "higher" species, and alternatives are to be used wherever possible. Supervision is carried out by a Federal Commission on Animal Experiments, comprised of experts, including animal welfare representatives, with ethical review of experiments carried out at a local level in each canton. The law provides that purpose-bred animals shall be used, although exemptions may be granted. There are also requirements for annual inspections and for records of each experiment to be kept for a period of 2 years.[27]

In December 1985, a national referendum was held on a proposed amendment to the constitution that would have banned most experiments involving live animals. The proposal was rejected by 70.5% of the voters.[28]

United Kingdom (1986)

It was noted in Chapter 1 that the 1876 Cruelty to Animals Act never satisfied the antivivisectionists, who saw it as a charter for vivisectors to do whatever they wanted. Over the century that the 1876 act was in force, there were several attempts to revise or reform it, not only by antivivisectionists, but also by reformers who merely wanted tighter controls. During the late 1960s and early 1970s, antivivisection groups, animal welfare organizations, and the newly emergent animal rights movement began another wave of activism in Britain. Terrorists broke into laboratories, sent letter-bombs to scientists, and bombed vehicles used to transport animals to laboratories. New questions were raised concerning the justification of research and the significance of the results. Continuing agitation for new legislation resulted in the introduction of several (unsuccessful) bills into Parliament. In 1979 Lord Halsbury, a former Chair of the Research Defence Society, introduced a private member's bill into the House of Lords. As a result of Lord Halsbury's concern, a Select Committee of the House of Lords was established, which included a number of eminent scientists. The committee report formed the basis for a prospective new law.[29]

The bill did not make much progress, but some of its provisions were noted by the government, which by this time was pledged to update the 1876 act. Under continued pressure by concerned groups, the Home Office asked the Animal Experimentation Advisory Committee to prepare a framework for yet another new law. By 1981 the framework was in place; by 1983 a White Paper was published to serve as a basis for revision and amendment. After much discussion among representatives of the government, scientists, and animal protection groups, a Supplementary White Paper was published in May 1985. This formed the basis of the 1986 bill, which passed through both Houses of Parliament by May 1986. A number of amendments were included that had been supported by various animal protection groups. Some of the provisions of the act were implemented on January 1, 1987; full implementation was to occur over the next 3 to 5 years.

An important aspect of the 1986 act is its establishment of a three-tier licensing system. The first level requires licensing of research, breeding, and supplying establishments (a "certificate of designation"), to ensure that they have adequate staffing and appropriate care for the animals. On the second tier, research workers carrying out technical work on animals will have to obtain a personal license (a "certificate of competence"). The third tier requires that the Secretary of State must carry out a cost–benefit analysis for every research program before a project license is granted.[30] A description of the British cost–benefit analysis reads as follows:

> The cost–benefit analysis weighs the benefits likely to accrue from the research against the predicted suffering of the animal. The moral principle is obvious and should result in research with high potential medical benefits and low animal suffering, whereas research causing high animal suffering and having low or zero potential benefit is unlikely to be permitted. The Secretary of State will be advised on this matter by an inspectorate (all of which have medical or veterinary qualifications) and by an animal procedures committee. Project license applications

request information about likely adverse effects of the procedures to be used, the way in which effects may be alleviated or avoided, whether alternatives have been considered and a suggested severity banding.[31]

Implicit in a cost–benefit analysis is the ability to assess severity by all involved in the research. *Severity* is the term used to encompass all descriptions of suffering, ie, pain, distress, discomfort, and other adverse effects on the animal. A scoring system is being developed to achieve more objectivity in assessing severity.[32]

At the time of this writing, it is too early to know how effective the new legislation will be in minimizing animal suffering and reducing the number of animals used in research.[33]

The European Convention for the Protection of Vertebrate Animals Used for Experimental and Other Scientific Research Purposes (Council of Europe)

This legislation, subject to approval by member countries, has been several years in development. Members approving the convention agree to register breeding, supplying, and user establishments and to use only purpose-bred animals for research. Covered animals are cats, dogs, and nonhuman primates (exceptions are permitted). The legislation stresses avoiding unnecessary suffering and ensuring competence of all personnel involved in the use of animals for research. It requires that veterinary consultation be available at each registered establishment. The convention establishes minimum standards for care and husbandry.[34]

The European Economic Community Directive for the Protection of Animals Used for Experimental and Other Scientific Purposes

This legislation is quite similar to that of the European convention and is also subject to approval by member countries. Individual countries that approve either of these laws may adopt stricter measures within their own country (as the United Kingdom and the Netherlands have done). The European Economic Directive outlines a commitment to work toward harmonizing regulations for testing across member states.[35]

Voluntary Regulation

Council of International Organizations of Medical Sciences (CIOMS)

The CIOMS is an international, nongovernmental scientific organization established jointly by the World Health Organization (WHO) and the United Nations Educational, Scientific and Cultural Organization

(UNESCO) in 1949. Its membership includes approximately 64 international organizations representing biomedical disciplines, and 26 national members representing medical academies and medical research councils. The main objective of the CIOMS is to foster international activities in the biomedical sciences. In December 1983, CIOMS held its XVIIth Round Table Conference for the purpose of developing guiding principles for biomedical research using animals. In 1985 the guiding principles were completed and made available to members, including a large number of representatives from developing countries that are just beginning to develop research programs.

The International Guiding Principles of the CIOMS are intended to provide a conceptual and ethical structure for whatever regulatory measures each country chooses to adopt with respect to animal use. The principles emphasize the importance of animal use in advancement of medical knowledge and improvement of health and well-being of both humans and animals. Alternative methods are encouraged wherever possible. When animals are used, the principles state that they should be appropriate for the study and should be used only after consideration of relevance to advancing knowledge or human and animal health. The principles further state that animals should be treated as sentient beings by all personnel, that minimization or avoidance of suffering should be regarded as ethical imperatives, and that investigators should assume that procedures that would cause pain in humans will also cause pain in other vertebrate animals. Investigators are required to provide appropriate anesthesia, analgesia, or sedation for procedures causing more than momentary pain, and the use of paralytic agents is discouraged (although waivers may be approved after suitable consideration and review). The principles also state that adequate housing facilities should be available for all species used for biomedical purposes, and their care should be under the supervision of a veterinarian. The director of the institute or department using animals is charged with the responsibility of assuring that all personnel are qualified for the procedures they are carrying out. Additional provisions include standards for transporting animals and for their housing and environmental conditions. There is also emphasis on nonanimal models such as computer techniques and the use of microorganisms.[36]

Canada

In Canada there is no federal legislation specifically for the regulation of the use of animals for research, although cruelty to animals is covered under Section 402 of the Criminal Code. There is, as well, some legislation concerning the transport of animals included in the Animal Disease and Protection Act. Three provinces in Canada have legislation covering aspects of animals in research (eg, Alberta has regulations for procurement of dogs and cats for research, teaching, and testing). Since 1968 the regulation of animal research throughout Canada has been carried out by the Canadian Council on Animal Care (CCAC), which is an autonomous advisory and supervisory body. The CCAC has implemented a system of self-regulation featuring local ethics committees as an alternative for legislation.

The CCAC requires that an ethics committee, or Animal Care Committee (ACC), be formed at each institution. The ACC must consist of senior scientific personnel who have experience with laboratory animals, including a veterinarian or biological scientist. The ACC is to be kept informed of all activities involving animal use and must establish procedures to ensure that, in any experiment likely to cause pain, the animal will be anesthetized or given analgesics, except when it would interfere with the purpose of the experiment. The ACC is responsible for training of personnel who care for animals. The ACC has the power to stop a procedure or to euthanize an animal to alleviate pain or distress if required procedures are not followed.[37]

The CCAC has published a two-volume *Guide to the Care and Use of Experimental Animals*, which is similar to the Public Health Service (PHS) guide in the United States. The guidelines of the CCAC have been recognized throughout the world, and parts of them are used in many countries. The CCAC has recently published *Ethics of Animal Experimentation*, which includes a set of principles for all those using vertebrates in research, teaching, or testing. Its *Categories of Invasiveness* relate to both vertebrates and cephalopods (octopuses and squid). Compliance with CCAC guidelines is a prerequisite for obtaining research funding. Continued failure to comply can result in withdrawal or withholding of major grants. The CCAC has also developed a *Syllabus of the Basic Principles of Laboratory Animal Science* (1985) for those using animals: research scientists, technicians, and other support personnel. The council has also produced a computer program for tracking animal use, entitled "Animal Research Protocol Management System."[38]

Australia

Legislation pertaining to animal experimentation has been passed in some Australian states. Within the past year, new or revised legislation relating to the use of animals in research and teaching has been passed through the Parliaments of New South Wales, South Australia, and Victoria. In each instance, the establishment of Animal Ethics Committees is required for institutions that will be licensed to conduct animal experiments. Representatives of community interests and of animal welfare will be required on the committees. There is at present no Commonwealth legislation relating to the welfare of animals used in experimentation. However, the issue of the use of animals for experimentation is presently under investigation by the Senate Select Committee on Animal Welfare.

The Australian Vice-Chancellors' Committee (AVCC) to the Senate Select Committee on Animal Welfare recently established a working party to achieve some resolution of this issue. In its submission to the Senate Select Committee on Animal Welfare, the AVCC supported a system of self-regulation by research and teaching institutions as the most effective way of ensuring the welfare of animals used in the conduct of teaching and research. The Vice-Chancellors' Committee has recommended that a system of review and monitoring, developed from the Canadian model, should be implemented. The committee also recommended the establishment of an autonomous Australian Council on Animal Care. In addition, the Australian Council on the Care of Animals in Research and Teaching (ACCART) was recently established to provide a national forum for communication among all interest groups. The Council is an independent, autonomous body whose primary role is advisory. Membership is drawn from government, research, and professional organizations as well as from the community.[39]

Israel

Despite the continuing political turmoil in Israel, there has been increasing public concern about the care and use of experimental animals. In response to these growing pressures, the government of Israel has begun to consider the needs of laboratory animals. In 1985 a special committee was appointed in Parliament (the Knesset) to review the problem and to seek some resolution. The committee recommended the adoption of the guidelines of the Council of International Organizations of Medical Sciences (CIOMS), modified to meet Israeli conditions, but the recommendation was rejected. A committee of representatives of Israeli medical schools reviewed the CIOMS guidelines and have now implemented them at their institutions. Researchers in Israeli medical schools will abide by the modified CIOMS guidelines unless or until the government or some other authority develops another approach.[40]

Other Countries[41]

Cyprus

Although Cyprus passed a Cruelty to Animals Law in 1910, it was a general law to protect animals with no provision for the use of animals in experimentation. However, provisions were made in the 1910 act for regulations regarding the transport of animals. Although there is currently no animal research legislation in Cyprus, experiments are carried out in line with the UK 1876 act. Formal permission is required from the Director of the Department of Veterinary Sciences and is given only to those with technical qualifications. Research is conducted mainly for vaccine testing and control.[42]

Portugal

Legislation is in preparation and will be in line with the Council of Europe Convention.[43]

Liechtenstein

As of this writing, Liechtenstein has not changed its 1936 prohibition on animal experimentation (see Chapter 1).

Notes

(The notation *[AB]* after a reference indicates that it is included in the Annotated Bibliography following Chapter 7.)

[1] Taylor GB (1975) Animal welfare legislation in Europe. In: *Animals and the law.* Hertfordshire, England: Universities Federation for Animal Welfare, pp 34–35.

[2] We thank Professor Thordal-Christensen for his translation of the laws and for arranging visits to research laboratories to demonstrate how the laws were implemented.

[3] Denmark, Act of Parliament—Statute 220, On Animal Experimentation (May 18, 1977).

[4] Regulation of animal use in selected foreign countries. (1986) In: Office of Technology Assessment *Alternatives to animal use in research, testing, and education.* Washington, DC: U.S. Government Printing Office, OTA-BA-273, pp 363–366.

[5] Radzikowski C (1984) Legislation in Poland. In: Bankowski Z, Howard-Jones N (eds) *Biomedical research involving animals.* (Proceedings of the XVIIth CIOMS Round Table Conference) Geneva: Council for International Organizations of Medical Sciences (CIOMS), p 140.

[6] Taylor GB (1975) Op cit, pp 37–38 [note 1]; Hampson J (1987) Law relating to animal experimentation. In: Tuffery AA (ed) *Laboratory animals: An introduction for new experimenters.* London: Wiley, p 48.

[7] Taylor GB (1975) Op cit, p 38 [note 1].

[8] Hampson J (1987) Op cit, p 49 [note 6].

[9] Finnish Animal Protection Act, No 91, Helsinki, Jan 27, 1971, and as amended by Act No 777 of Sept 30, 1985. Ormala H, Nevalainen T (1982) Pertinent laboratory animal legislation in Finland. In: Pelkoneu K, Nevalainen T, Hänninen O (eds) *Laboratory animals: Use, legislation, and future.* Kuopio: Kuopion korkeakoulu, pp 11–13; Lahteensmaki V (1987) Legislation dealing with animal experimentation in Finland. *An Tech 38*,3:229–233. We thank Professor Osmo Hänninen of the University of Kuopio for providing many of these details in personal discussions with J.A.S. We also thank Professor Heikki Päivärinta of the University of Helsinki for providing information about the Finnish Act and its 1985 amendments.

[10] Hampson J (1987) Op cit, p 48 [note 6]; Federal Republic of Germany, Animal Protection Act of July 24, 1972, Sartorius Band I, Verfassungs- und Verwaltungsgesetze, 873, 1–12, Ergänzungslieferung Marz (1982); Federal Republic of

Germany, Federal Ministry of Food, Agriculture, and Forestry, List of Regulations Affecting the Conduct of Animal Experiments, BML No 321–3520–1/1 (Bonn: Oct 17, 1984); Federal Republic of Germany, Federal Ministry of Food, Agriculture and Forestry, Proposal to the Cabinet for Amendments to the Animal Protection Act of July 24, 1972 (Bonn: Nov 1984); Regulation of animal use in selected foreign countries. (1986) Op cit, pp 366–367 [note 4].

11 German Law on Animal Protection, Aug 18, 1986, and Amendments, Bundesgesetzblatt, Teil 1, Section I-XII, implemented Jan 1, 1987. In 1988 additional amendments were passed. Among them were new recordkeeping requirements for all investigators using vertebrates in research. (Bundesgesetzblatt Teil 1, Nr 37 - Tag der Ausgabe: Bonn, Aug 4, 1988.)

12 Hampson J (1987) Op cit, p 49 [note 6].

13 Japan, Prime Minister's Office The law and related standards for the protection of animals. (Law No 105, Oct 1, 1973); see also Regulation of animal use in selected foreign countries. (1986) Op cit, pp 362–363 [note 4]. We are grateful to Dr. Chizuko Yamaguchi at the Japan Animal Welfare Society (JAWS) in Tokyo for providing information about the new Japanese law, which, at the time of this writing, had not yet been translated into English.

14 Taylor GB (1975) Op cit, p 34 [note 1]; Hampson J (1987) Op cit, p 47 [note 6].

15 Taylor GB (1975) Op cit, p 39 [note 1].

16 Hampson J (1987) Op cit, pp 50–51 [note 6].

17 Regulation of animal use in selected foreign countries. (1986) Op cit, pp 368–369 [note 4]. We are indebted to Professor Stian Erichsen at the National Institute of Public Health in Oslo for providing many of these details during a visit by J.A.S. to Oslo in 1985.

18 Taylor GB (1975) Op cit, p 34 [note 1].

19 Hampson J (1987) Op cit, p 47 [note 6].

20 Taylor GB (1975) Op cit, pp 38–39 [note 1].

21 Hampson J (1987) Op cit, p 50 [note 6].

22 Regulation of animal use in selected foreign countries. (1986) Op cit, pp 367–368 [note 4].

23 Hampson J (1987) Op cit, p 51 [note 6]; Regulation of animal use in selected foreign countries. (1986) Op cit, pp 369–371 [note 4].

24 We appreciate the many communications from Professor Karl Öbrink at the University of Uppsala in Sweden concerning the progress of the Swedish system.

25 Gribetz S (1989) Newsfronts: Farm animals of Sweden gain rights. *Lab Anim* (NY) *18*,1:8–9.

26 Taylor GB (1975) Op cit, p 42 [note 1].

27 Hampson J (1987) Op cit, p 52 [note 6]; Regulation of animal use in selected foreign countries. (1986) Op cit, pp 371–372 [note 4]; Swiss Academy of Medical Sciences and Swiss Academy of Sciences (1983) *Principles and guide-lines for scientific experiments on animals.*

28 Swiss reject plan to bar animal experiments (Dec 2, 1985) *New York Times.*

29 Hampson J (1987) Op cit, p 23 [note 6].

30 United Kingdom, Animals (Scientific Procedures) Act 1986, Elizabeth II, Ch 14 (May 20, 1986); Hampson J (1987) Op cit, pp 26–30 [note 6]; Morton DB (1987) Epilogue: Summarization of colloquium highlights from an international perspective. *J Am Vet Med Assoc 191*,10:1294–1296.

[31] Morton, ibid.
[32] Ibid.
[33] We thank Dr. David B. Morton, University of Leicester, for explaining many details of the new British law.
[34] Morton DB (1987) Op cit [note 29]; International agreements governing animal use. (1986) In: Office of Technology Assessment, op cit, pp 412–414 [note 4].
[35] Morton, ibid.
[36] Howard-Jones N (1985) A CIOMS ethical code for animal experimentation. *WHO Chronicle 39*:51–56. At the time of this writing, the CIOMS Guiding Principles have not been adopted by the WHO as the Human Experimentation Code was.
[37] Flowers FH (1983) Research animal care in Canada: Its control and regulation. In: Sechzer JA (ed) *The role of animals in biomedical research.* Annals of The New York Academy of Sciences, vol 406. New York: The New York Academy of Sciences, pp 144–149 [AB]; Rowsell HC (1984) Monitoring of animal experimentation: Codes of practice. In: Bankowski Z, Howard-Jones N (eds), op cit, pp 145–155 [note 5].
[38] We thank Dr. Harry C. Rowsell, Director of the Canadian Council for Animal Care, for bringing us up-to-date on the Canadian regulations and guidelines.
[39] This information was provided by Dr. Margaret Rose, University of New South Wales, New South Wales, Australia. In 1989 The Australian Council on the Care of Animals in Research and Testing (ACCART) started publishing a newsletter to inform scientists and the public about issues concerning animal use.
[40] Information about activities in Israel concerning laboratory animal use was provided by Professors Jonathan Adler and Shaul Feldman, Hebrew University Medical School in Jerusalem, at a meeting held with J.A.S. in 1987.
[41] We have included information in this chapter on all countries for which information was available. In omitting any country, we do not mean to imply that it has no legislation or other form of regulations pertaining to animal research.
[42] Hampson J (1987) Op cit, p 47 [note 6]; Taylor GB (1975) Op cit, p 34 [note 1].
[43] Hampson J (1987) Op cit, p 51 [note 6].

Annotated Bibliography

1966

Articles

American Association for the History of Medicine (1966) Resolution adopted by the American Association for the History of Medicine at its 39th annual meeting at Rochester, Minnesota, May 14, 1966. *Bull Hist Med 40*,3:277–278.
The resolution opposes all legislative restrictions on animal research (while supporting legislation "to control the procurement, sale and care of laboratory animals") and emphasizes the many contributions of animal experiments to medical knowledge.

Burchell HB (1966) Dogs and the search for health. *Circ 34*,2:185–187.
Dogs have played an important role in the study of cardiovascular disease, but this role is threatened by restrictive legislation. Explains the position of the National Society for Medical Research (NSMR) in endorsing legislation related to the procurement of animals and their housing and care; however, the NSMR opposes restrictive legislation.

Dennis C (1966) America's Littlewood crisis. The sentimental threat to animal research. *Surg 60*,4:827–839.
Reviews the history of antivivisectionism in Great Britain and the United States and discusses bills introduced in Congress over the past several years. The author urges cardiovascular surgeons to take a more active part in fighting restrictions, and he deplores "the rabid nature of our opposition."

Freeman A (1966) Law to protect laboratory animals. *J Am Vet Med Assoc 149*,10:1330–1332.
Describes the background of the Laboratory Animal Welfare Act (PL 89–544) and some of its provisions. The author is wary of the new law, which had just been passed, but he concludes that "about all that can be done now is to adopt a wait-and-see attitude, for the bill is now law."

Hirschboeck JS (1966) Animals and human welfare. *Hosp Progr 47*, 5:98–99, 122.
A response to an antivivisectionist article circulated by the National Catholic Society for Animal Welfare. The author objects to identifying Catholicism with antivivisectionism, which is "inimical to human welfare." He endorses a statement by the National Society for Medical Research, reproduced in full, which opposes restrictive legislation.

Ingle DJ (1966) Scientific and ethical responsibilities associated with the use and care of animals. *Fed Proc 25*,5:1465–1467.
Ethical responsibilities of the scientist include procuring animals legally, controlling disease in the animal facility, and not causing any unnecessary pain. Agonizing pain cannot be justified, but humaneness can be achieved only through education. The author believes that humaneness cannot be achieved through legislation.

Langer E (1966) Animal care: The humane movement is pulling ahead. *Science 151*:1515–1518.
Contains detailed descriptions of the provisions of various bills under consideration and the constituencies of each. Concludes that, even if some legislation passes, the issue "is likely to stay with us for a long while."

Locke RF (1966) Legislation and research. *Ill Dent J 35*,6:331–332.
The author, a veterinarian, opposes the proposed Laboratory Animal Welfare Act, which he believes is a threat to research. He argues that animals are already well cared for. Certification of facilities is recommended instead of legislation.

Ozer MN (1966) The British vivisection controversy. *Bull Hist Med 40*:158–167.
Discusses the antivivisection movement in 19th century England and political events leading to the passage of the Cruelty to Animals Act of 1876. The author describes the act as "an example of how medicine may be affected by . . . the anti-intellectual *extremism* of any period."

Taussig HB (1966) Animal legislation and our program. Public Law 89-544. *Circ 34*,6:1114–1116 (also published in *Circ Res* [1966] *19*,6:1110–1112).
States the provisions of the recently enacted Laboratory Animal Welfare Act and urges researchers to make every effort to comply. Advocates increased use of pound animals to offset costs of compliance; endorses accreditation for animal facilities by the American Association for the Accreditation of Laboratory Animal Care (AAALAC).

Visscher MB (1966) The Congress and animal experimentation. *J Am Med Assoc 196*,12:1053–1054.
Written as Congress was considering legislation that eventually passed as the Laboratory Animal Welfare Act of 1966, this article strongly opposes

all restrictive legislation, especially legislation that would regulate any part of the experimental design.

Wolfle D (1966) Animal care legislation. [editorial] *Science 153*:1061.
Indicates resigned acceptance of the Laboratory Animal Welfare Act and warns that animal welfare groups will now push for more stringent regulations unless scientists convince Congress that they can regulate themselves through the American Association for the Accreditation of Laboratory Animal Care (AAALAC).

1967

Book

Roberts C (1967) *The scientific conscience: Reflections on the modern biologist and humanism.* New York: George Braziller (published in Fontwell, Sussex, by Centaur Press [1974]).
The chapters were written as separate essays; four were previously published. The theme uniting all the essays is a critique of modern science, especially biological science (the author is a microbiologist), as dehumanizing. Two chapters address animal experimentation specifically and find it largely unjustified.

Articles

Jones EM (1967) The Laboratory Animal Welfare Act (PL 89–544). *J Am Vet Med Assoc 151*,12:1862–1866.
A USDA veterinarian explains the background of the act and the steps being taken by the USDA to implement it. Some specific requirements are discussed, for example, cage sizes.

McCann MB, Stare FJ (1967) The contribution of animal studies to nutritional discoveries that have benefitted both animal and man. *Am J Pub Health 57*,9:1597–1604.
Paper presented at a symposium entitled "Vivisection—Vivistudy: The facts and the benefits to animal and human health." Documents the knowledge about nutrition and deficiency diseases that have come from animal studies.

Melby EC (1967) Problems in animal procurement. *J Surg Res 7*,6:294–298.
Examines the effect of the 1966 Laboratory Animal Welfare Act on procurement of animals for research. Argues for more use of pound animals, but acknowledges that this practice can be made acceptable to the public only if animal facilities and care are improved.

Robinson HJ (1967) Animal experimentation leading to the development of drugs benefitting human beings and animals. *Am J Pub Health 57*,9:1613–1620.
Paper presented at a symposium entitled "Vivisection—Vivistudy: The facts and the benefits to animal and human health." Describes "the many and varied ways in which animal experimentation served in the development of drugs," noting that many of these drugs are beneficial to animals as well as humans.

Schiller J (1967) Claude Bernard and vivisection. *J Hist Med 22*:246–260.
A tribute to Claude Bernard and the contributions of vivisection to scientific progress, especially to knowledge of the nervous system. About two pages are devoted to antivivisectionist protests of the 19th century.

Soave OA (1967) Animal experimentation leading to better care of laboratory and pet animals. *Am J Pub Health 57*,9:1621–1626.
Paper presented at a symposium entitled "Vivisection—Vivistudy: The facts and the benefits to animal and human health." The author states that "the purpose of this paper is to explore the question, 'Have the animals themselves gained because of animal research and what have they benefited through man's exploitation of them?'" The question is answered affirmatively, and the paper discusses several ways in which animals have profited from research on other animals.

Visscher MB (1967) Medical research and ethics. *J Am Med Assoc 199*,9:631–636.
Opposes legislation currently before Congress because these bills are antivivisectionist in nature, would impede science and teaching, and place animal welfare ahead of human welfare. Self-regulation of animal research, not government regulation, is recommended.

Winterscheid LC (1967) Animal experimentation leading to the development of advanced surgical techniques. *Am J Pub Health 57*,9:1604–1612.
Paper presented at a symposium entitled "Vivisection—Vivistudy: The facts and the benefits to animal and human health." Documents the contributions of animal experiments to three areas of surgery: hemorrhagic shock, cardiac lesions, and organ transplants.

1968

Articles

American Psychological Association. APA Board of Scientific Affairs (1968) Principles for the care and use of animals. *Am Psychol 23*,11:819.
List of nine principles approved June 26, 1968 by the Council of Representatives of the American Psychological Association.

American Society for Artificial Internal Organs (1968) ASAIO statement concerning the use of animals for research in the field of artificial organs. *Trans Am Soc Artif Intern Organs 14*:454.
A five-point statement approved April 21, 1968, by the ASAIO. Affirms the need for animal experimentation and for high standards of care; urges compliance with the Animal Welfare Act; calls for funding with which to upgrade facilities; calls for educating the public about the use of animals in research.

Editorial notes—Laboratory animal care. [unsigned editorial] (1968) *Hospitals 42*,23:35.
Applauds the efforts of the American Association for the Accreditation of Laboratory Animal Care (AAALAC), founded 3 years earlier, to provide effective voluntary regulation of laboratory animal facilities. Urges hospitals with animal research facilities to apply for AAALAC accreditation.

Jones EM (1968) The Laboratory Animal Welfare Act: Regulations, standards, enforcement, and progress. *J Am Vet Med Assoc 153*,12:1874–1887.
A senior staff veterinarian with the USDA explains how the regulations and standards to enforce the act were developed. Reports that the act has resulted in improvements "even in actual research, which is not covered under this relatively new legislation."

Kanterman CB (1968) A dirty word called "vivisection." [editorial] *Dent Stud 47*,3:212–213.
This piece was inspired by the issuance of a postage stamp honoring John Dewey, who defended the use of animals in research. The author quotes Dewey's justification for animal research and comments that, although "vivisection has acquired a disreputable connotation," the public has finally come to realize its benefits.

1969

Articles

Breazile JE, Kitchell RL (1969) Pain perception in animals. *Fed Proc 28*,4,pt.2:1379–1382.
The volume in which this appears was published by the Veterans Administration as a *Technical guide for comparative anesthesia in laboratory animals*; this paper is the second of 32 chapters dealing with various aspects of anesthesia. Explains similarities and differences in the neurophysiology of pain perception in animals and humans; takes the position that pain perception in animals is analogous to pain perception in humans.

Kaplan H (1969) Anesthesia in invertebrates. *Fed Proc 28*,4,pt.2:1557–1569.
This is Chapter 30 in a volume published by the Veterans Administration, intended as a *Technical guide for comparative anesthesia in laboratory animals.* Contains an extensive table listing information about appropriate anesthetics for 10 phyla of invertebrates. The author believes that the investigator should use anesthetics for "all multicellular forms of macroscopic dimensions."

Price DA (1969) Experimental use of animals by secondary school students. *J Am Vet Med Assoc 155*,7:1264–1265.
The American Veterinary Medical Association endorses the National Society for Medical Research's (NSMR) "Guiding Principles in the Use of Animals by Secondary School Students and Science Club Members" adopted by the NSMR in June. The guidelines are printed in full and discussed.

1970

Articles

Benison S (1970) In defense of medical research. *Harv Med Alumni Bull 44*,3:16–23.
A historical account of skirmishes between the medical research community and antivivisectionists in the United States between 1866 (when the ASPCA was founded) and the 1920s. Focuses on activities of Harvard professor Walter B. Cannon to fight restrictive legislation. Includes a reprint of Cannon's 1909 code for laboratory procedures.

Bustad LK (1970) The experimental subject: A choice not an echo. *Perspec Biol & Med 14*,1:1–10.
The author, who played a role in the development of a strain of miniature pigs for research, urges scientists to consider using not only pigs, but other animals not usually found in laboratories: especially bats, armadillos, and marmosets. However, all animals should be treated with sensitivity, and not "as if they were things."

Decker WM (1970) Laboratory animal care legislation. *J Am Vet Med Assoc 157*,3:284–285.
A statement presented in testimony at the Public Hearings on Animal Humane Care and Housing Bill (the Whitehurst amendment to the Animal Welfare Act) held June 8. The author, representing the American Veterinary Medical Association (AVMA), testified in opposition to two key provisions of the bill: (1) the proposal to extend coverage to all warm-blooded animals, and (2) the proposal to extend coverage to the research itself. [Note that the latter provision was dropped from the final version

of the bill that became law, but coverage was extended to include "such other warm-blooded animal, as the Secretary (of Agriculture) may determine is being used, or is intended for use, for research, testing, experimentation purposes, or as a pet."]

Saulmon EE (1970) Implementation of laboratory animal laws. *J Am Vet Med Assoc 157*,11:1964–1968.
A USDA official examines the background of the Animal Welfare Act, its significance, and its reception in the scientific community. Describes how the USDA is organized to carry out inspections. Discusses briefly some of the 14 bills currently in Congress relating to laboratory animal welfare.

1971

Articles

Anonymous (1971) Antivivisection rides again. [editorial] *J Am Med Assoc 217*,1:70.
Argues that "the use of animals for experimental purposes is essential to biomedical research and teaching." Discusses several recent developments, mostly legislative, that threaten animal research.

Eccles JC (1971) Animal experimentation versus human experimentation. In: National Research Council. Institute of Laboratory Animal Resources, *Defining the laboratory animal.* Washington, DC: National Academy of Sciences, pp 285–293.
The author's objective is to stress that the ethical restrictions on using humans for experimentation require increased use of animals. He sees no ethical restrictions that would apply to animals, and he believes there is general agreement about this.

Fox MW (1971) Effects of rearing conditions on the behavior of laboratory animals. In: National Research Council. Institute of Laboratory Animal Resources, *Defining the laboratory animal.* Washington, DC: National Academy of Sciences, pp 294–309.
Discussion of how environmental factors can affect the behavior of laboratory animals and, therefore, experimental results. The humaneness of artificial, impoverished laboratory environments is questioned, especially for wild-born primates. Urges researchers not to forget "that all mammals have a 'psyche' as well as a physiology, and ignorance is no excuse for inhumane treatment."

Visscher MB (1971) The Animal Welfare Act of 1970 (PL 91–579). *Science 172*:916–917.
The president of the National Society for Medical Research (NSMR) discusses how the 1970 act differs from the one passed in 1966. Because

of the 1970 act's "specific prohibition against any interference by the Secretary of Agriculture with the design or execution of any experiment," the NSMR endorsed it.

1972

Articles

American Psychological Association. APA Committee on Precautions and Standards in Animal Experimentation (1972) Guidelines for the use of animals in school science behavior projects. *Am Psychol 27*,4:337.
List of nine guidelines to govern the use of animals in secondary schools. Introductory paragraph notes the need for such guidelines because of the growing use of live animals in classroom science projects. These guidelines were revised in 1981 (see American Psychological Association, 1981).

Finch CO (1972) Trends in humane animal care legislation. *J Am Vet Med Assoc 161*,11:1533–1535.
Animal welfare legislation in the United States from the 17th century to the present is reviewed to support the theme that many people are now becoming aware of the issues. The author, a veterinarian, believes that all veterinarians should welcome this trend "to assure healthier animals spared the cruelties of the past."

Smith CR, Booth NH, Fox MW, et al (1972) Report of the AVMA panel on euthanasia. *J Am Vet Med Assoc 160*:761–772.
This was the second of four euthanasia reports published to date by the American Veterinary Medical Association (1963, 1972, 1978, and 1986). It differs from the 1963 version (*J Am Vet Med Assoc 142*:162–170) in evaluating a few more methods of euthanasia, including some used specifically in research laboratories (the earlier report dealt only with methods used in pounds and animal shelters). The 1963 condemnation of some methods is softened in this report, but most recommendations remain substantially the same. (For the third revision, see McDonald LE, et al, 1978. The 1986 revision was published in *J Am Vet Med Assoc 188*,3:252–268; an abbreviated version of it is reprinted in the Interdisciplinary Principles and Guidelines reproduced in this book in Appendix C.)

1973

Article

Visscher MB (1973) Commentary on laws relating to the scientific use of unclaimed impounded animals. *J Am Vet Med Assoc 163*,1:78–79.
A table is presented summarizing the author's research on state and local laws pertaining to the release of pound animals for scientific study. Ten

states have laws requiring the release of unclaimed, impounded animals to research institutions. The author argues that "it would be desirable if every state had [such] legislation."

1974

Articles

Best CH (1974) A short essay on the importance of dogs in medical research. *Physiol 17*,4:437–440.
The author describes his work with Fred Banting conducting experiments on dogs that eventually led to the discovery of insulin as a treatment for diabetes. An editor's note states that the article was written "because of the recent great interest in the Congress of the United States in restricting or even prohibiting the use of dogs in biomedical research." Best declares that he and Banting never "inflicted pain and discomfort on dogs which we would not have been willing to accept ourselves."

Holden C (1974) Beagles: Army under attack for research at Edgewood. *Science 185*:130–131.
News feature about the furor over the Department of Defense's (DOD) use of beagles in toxic substance experiments. The army has suspended its beagle research pending review by the DOD. A bill has now been introduced in Congress that would prohibit the military from using dogs in research related to chemical warfare.

1975

Book

Visscher M (1975) *Ethical constraints and imperatives in medical research.* Springfield, IL: CC Thomas.
Two chapters deal with animal experimentation. Chapter 11 ("The ethics of the use of lower animals in scientific study") defends animal research, and Chapter 12 ("Animal experimentation and the law in relation to ethics") argues that scientific research should be free of legal restrictions.

Articles

Mikeska JA, Klemm WR (1975) EEG evaluation of humaneness of asphyxia and decapitation euthanasia of the laboratory rat. *Lab Anim Sci 25*,2:175–179.
Experiments were done to assess the relative humaneness of killing rats by asphyxia (accomplished by turning off the respirator of paralyzed, conscious rats) and decapitation (by guillotine). The author concluded that the EEG monitoring indicated that the rats suffered in both cases:

for a longer time with asphyxia, but more intensely with decapitation. This report formed the basis for The American Veterinary Medical Association's recommendation against decapitation without prior sedation contained in the AVMA's 1986 euthanasia guidelines (Smith AW, et al [1986] 1986 report of the AVMA panel on euthanasia. *J Am Vet Med Assoc 188*,3:252–268; an abbreviated version of this report is reprinted in this book as an addendum to Appendix C).

Powledge T (1975) "Primate"—Sensationalism for subtlety. *Hastings Ctr Rep 5*,1:5–6.
A critical review of *Primate*, the documentary film by Frederick Wiseman, filmed at the Yerkes Regional Primate Research Center in Atlanta. The author accuses Wiseman of choosing "the easy targets, the cheap shots, the gory and sensational."

Thomas GJ (1975) Editorial. *J Comp & Physiol Psychol 89*,1:1–4.
This long piece is concerned mostly with unrelated matters of editorial policy, but it is notable for one section on animal experimentation in which several specific procedures are proscribed. Warns that failure to comply is grounds for rejecting papers for publication.

WGBH Public Television (1975) Primate and the scientific ethos. Can some knowledge simply cost too much? *Hastings Ctr Rep 5*,1:6–8.
This is a much abbreviated and edited transcript of a panel discussion entitled *The price of knowledge?* produced by WGBH in Boston and broadcast on public television channels. The panelists discuss the documentary film *Primate* made by Frederick Wiseman.

1976

Book

Pratt D (1976) *Painful experiments on animals.* New York: Argus Archives.
A highly critical discussion of the pain and suffering endured by experimental animals. Examples of painful experiments excerpted from the scientific literature are mostly from New York State. Includes information on the statistics of animal use, laws, and alternative methods. (For the sequel to this book, see Pratt D, 1980.)

Articles

Aledort LM, Grupenhoff JT (1976) Legislative considerations for studies with animal models. In: National Institutes of Health *Animal models of thrombosis and hemorrhagic diseases.* Bethesda, MD: National Institutes of Health, pp 159–165.
Discusses administrative problems in implementing USDA regulations under the Animal Welfare Act, since Congress has allocated very little money for this purpose.

Anchel M (1976) Beyond "adequate veterinary care." *J Am Vet Med Assoc 168*,6:513–517.
The requirement of the Animal Welfare Act that laboratory animals be given "adequate veterinary care" ought to mean more than just physical health. The author argues that it means also "the maintenance of a healthy emotional state."

Hull D (1976) The rights of animals. [book review] *Science 192*:679–680.
Review of Singer P (1975) *Animal liberation* (New York: New York Review/Random House). The reviewer, a philosopher, finds Singer's argument for animal liberation and against speciesism to be unconvincing. He objects to what he takes to be the gist of the book, that "we owe other animals so much, and they owe us nothing."

Leeper EM (1976) NAS symposium: New methods may reduce needs, can't replace laboratory animals. *BioSci 26*,1:9–12.
Summary of the National Academy of Sciences symposium entitled "The Future of Animals, Cells, Models and Systems in Research, Development, Education, and Testing." Some concrete suggestions emerging from the symposium are listed.

Melby EC Jr (1976) What does "adequate veterinary care" mean to the American College of Laboratory Animal Medicine? *J Am Vet Med Assoc 168*,6:508–509.
Defends animal research and standards of animal care in the United States. The author believes that "adequate veterinary care" is adequately defined in the Animal Welfare Act, but the problem is in finding enough trained veterinarians and in giving them enough authority to provide this care.

Meth T (1976) Legal implications in the alternatives to the use of experimental animals. *J Clin Pharmacol 16*,10,pt 2:589–591.
A warning from a lawyer to the biomedical community, particularly pharmacologists, that they will end up rigidly hampered by legislation unless they "exercise leadership in the liberation of other species from the pain and death so often involved in experimentation—particularly under FDA directives—and in the aggressive search for funds to develop the alternatives to the use of animals."

Pakes SP (1976) Adequate veterinary care as viewed by the American Association for Accreditation of Laboratory Animal Care. *J Am Vet Med Assoc 168*,6:519–521.
Describes the requirements an animal facility must meet in order to qualify for accreditation by the American Association for Accreditation of Laboratory Animal Care (AAALAC); reports that about 280 institutions to date are accredited. The author believes that humane care is provided in most cases, although some differences of opinion exist as to

what humane care is; "those differences will always be a stimulus for discussion."

Phalen RF (1976) Inhalation exposure of animals. *Env Health Persp 16*(Aug):17–24.
Technical discussion of various inhalation exposure methods, advantages and disadvantages of each, and design criteria for each. Humane considerations are mentioned throughout, followed by a concluding section on "the ethics involved in inhalation exposures."

Schwindaman DF (1976) An evaluation of adequate veterinary care by the US Department of Agriculture. *J Am Vet Med Assoc 168*,6:517–519.
Discusses the responsibilities placed on veterinarians by the 1970 amendments to the Animal Welfare Act that include provisions for pain relief under the definition of "adequate veterinary care," and the requirement that the use of pain-relieving drugs be included in the facility's annual reports to the United States Department of Agriculture. Sections of USDA standards and Veterinary Services Memorandum pertaining to pain relief are reproduced in full.

Stevens C (1976) Humane considerations for animal models. In: National Institutes of Health *Animal models of thrombosis and hemorrhagic diseases*. Bethesda, MD: National Institutes of Health, pp 151–158.
The paper deals with "the humane care of animals as it relates to the specific study of spontaneous animal disease," and not experimentally induced conditions. Suggestions include providing an enriched environment and providing for privacy, sociability, and exercise as different species require.

Vierck CJ Jr (1976) Extrapolations from the pain research literature to problems of adequate veterinary care. *J Am Vet Med Assoc 168*,6:510–513.
A general discussion of how little is known about animal pain and suffering—defined as the emotional reaction to pain—and the need to learn more. The author argues that there is enough behavioral evidence to warrant the assumption that "these animals not only feel pain but suffer from it." Includes several suggestions for studying pain humanely. Opposes legislation and the establishment of guidelines.

Wade N (1976) Animal rights: NIH cat sex study brings grief to New York museum. *Science 194*:162–167.
A lengthy report on the issues involved in protests by animal welfare groups against a study of the sexual behavior of cats at the American Museum of Natural History. Both sides of the controversy are examined. The reporter found the charge of poor care for the cats to be groundless, but he concludes the piece on a note that is critical of current scientific codes of ethics. This article inspired a lively dialogue in numerous letters over the next several issues.

1977

Book

National Research Council-National Academy of Sciences. Institute of Laboratory Animal Resources (1977) *The future of animals, cells, models, and systems in research, development, education, and testing*. Washington, DC: National Academy of Sciences.
Proceedings of a symposium held in October 1975. Animal welfare activists and scientists addressed questions of ethics, legislative controls, the contributions of animals in research, and possibilities for alternative methods. Three papers are directed primarily to ethical issues (see Hummer RL, 1977; Stevens C, 1977; Rowsell HC, 1977), and two papers address legislative questions (see Foley TS, 1977; Ladimer I, 1977).

Articles

Bustad LK, Hegreberg GA, Padgett GA (1977) Animal models. In: National Research Council. Institute of Laboratory Animal Resources, *The future of animals, cells, models, and systems in research, development, education, and testing*. Washington, DC: National Academy of Sciences, pp 130–149.
A review of naturally occurring animal models for genetic disorders and congenital malformations, chronic degenerative diseases, and cancer. The authors see the need for more animals for research in the future rather than less. A brief concluding section, "The Philosophy of Animal Research," discusses the investigator's moral responsibilities in terms of Russell and Burch's "three Rs" (replacement of animals, reduction in numbers of animals needed, and refinement of techniques to reduce suffering).

Davison FC (1977) Historical perspectives of biomedical experimentation. In: National Research Council. Institute of Laboratory Animal Resources, *The future of animals, cells, models, and systems in research, development, education, and testing*. Washington, DC: National Academy of Sciences, pp 7–15.
A catalog of some important contributions made to biomedical knowledge through brief biographical sketches of scientists who used animals in their research, from Aristotle and Galen to Banting and Best. The author draws several conclusions from history, all of which affirm the need for continued use of animals in teaching and research.

Foley TS (1977) A legislator's view on animal legislation affecting biomedical research. In: National Research Council. Institute of Laboratory Animal Resources, *The future of animals, cells, models, and systems in research, development, education, and testing*. Washington, DC: National Academy of Sciences, pp 171–175.

The author, a representative from the state of Washington, describes legislation currently before Congress and discusses the problem of "pitifully inadequate" funding appropriated for the enforcement of animal welfare laws.

Goodrich JE (1977) The first 100 years of antivivisection 1824–1924. *Mayo Clin Proc 52*:257–259.
Brief chronological account of antivivisectionist activities in England and the United States from 1824 (the date of the founding of the Royal Society for the Prevention of Cruelty to Animals) to 1924 (shortly after a bill was defeated that would have prohibited all experiments on living dogs in the United States). The history of antivivisection is characterized as the story of "those who misunderstand the scientific methods, in spite of the facts that are presented to them."

Harrell GT (1977) Introduction. In: National Research Council. Institute of Laboratory Animal Resources, *The future of animals, cells, models, and systems in research, development, education, and testing.* Washington, DC: National Academy of Sciences, pp 1–6.
Introduction to the symposium proceedings, "The Future of Animals, Cells, Models and Systems in Research, Development, Education, and Testing." Emphasizes medical advances made in this century through animal research and the need to pay attention to animal welfare for the sake of scientific validity.

Hummer RL (1977) The era of humane awareness. In: National Research Council. Institute of Laboratory Animal Resources, *The future of animals, cells, models, and systems in research, development, education, and testing.* Washington, DC: National Academy of Sciences, pp 25–33.
The author presents the views of the members of the American Humane Association (AHA). The AHA is not antivivisectionist but takes the position that some researchers do expose animals to unnecessary pain because of inadequate training and a lack of compassion.

Ladimer I (1977) Root and branch: Legal aspects of biomedical studies in man and other animals. In: National Research Council. Institute of Laboratory Animal Resources, *The future of animals, cells, models, and systems in research, development, education, and testing.* Washington, DC: National Academy of Sciences, pp 296–314.
Discusses legal bases for statutes regarding animal experimentation, and the need for continued animal research. The author argues that no more law in this area is needed; scientists should emphasize professional self-regulation instead. Suggests establishing a legal information and advisory service for researchers.

Musto DF (1977) Freedom of inquiry and subjects' rights: Historical perspective. *Am J Psychiatry 134*,8:893–896.

This paper was presented at a symposium, "Freedom of Inquiry and Subjects' Rights," at the 1976 meeting of the American Psychiatric Association; it was the only paper to deal with animal, as well as human, experimentation. A historical sketch of attitudes toward in vivo experimentation is presented, contrasting the "antiscientific sentimentality" of GB Shaw, who was an antivivisectionist, with the views of WH Welch, first Dean of Johns Hopkins Medical School. Welch testified in favor of animal research at Senate hearings in 1900. The author sees similarities between "present criticism of research on humans [and] the heyday of antivivisection activity in the last century;" both reflect a questioning of the value of unlimited progress if it means risks to the quality of life.

Pattishall EG Jr (1977) Animal behavior: Relation to illness and disease. In: National Research Council. Institute of Laboratory Animal Resources, *The future of animals, cells, models, and systems in research, development, education, and testing*. Washington, DC: National Academy of Sciences, pp 115–123.
A justification of animal experimentation in behavioral research on the grounds that behavioral components (psychological and social) have important effects on disease and illness. The author considers the dilemma posed by stress and deprivation studies, but concludes that they are justifiable.

Rowsell HC (1977) The ethics of biomedical experimentation. In: National Research Council. Institute of Laboratory Animal Resources, *The future of animals, cells, models, and systems in research, development, education, and testing*. Washington, DC: National Academy of Sciences, pp 267–281.
Dismissing extremist positions on both sides of the controversy, the author urges "intelligent, rather than emotional, solutions." Justifies the use of animals in biomedical research and testing, but emphasizes that each project should be examined carefully, and that Russell and Burch's "three Rs" should be applied. The specific responsibilities of the scientist, the director of animal care, the technicians, and the institution are discussed.

Schneider HA (1977) Summary. In: National Research Council. Institute of Laboratory Animal Resources, *The future of animals, cells, models, and systems in research, development, education, and testing*. Washington, DC: National Academy of Sciences, pp 318–327.
Summarizes symposium papers. Many speakers emphasized that there is no substitute for the intact animal in much research, but the new in vitro methods being developed are useful adjuncts. Justifies animal research in terms of the Judeo-Christian ethic, which tells us that "man is above the animals."

Smith RJ (1977) Creative penmanship in animal testing prompts FDA controls. *Science 198*:1227–1229.

This article was written upon completion of the FDA's first round of revisions of the Good Laboratory Practice regulations (GLPs), which had initially been proposed in 1976. Provides a good account of the sloppy laboratory practices discovered by the FDA that led to the development of the regulations. (The author states that the GLPs were to go into effect in January 1978, but they underwent further revisions and were not published in their final form until December 1978; they went into effect in June 1979.)

Stevens CG (1977) Humane perspectives. In: National Research Council. Institute of Laboratory Animal Resources, *The future of animals, cells, models, and systems in research, development, education, and testing.* Washington, DC: National Academy of Sciences, pp 16–24.
The author hopes that recent developments signal a change "from the early 1960s, which saw such flat, solid, seemingly unbreakable opposition in the American scientific community to mandatory humane requirements." Laboratory animal treatment is criticized, and scientists are urged to improve conditions for laboratory animals to avoid sharpening the confrontation with animal welfare groups.

1978

Book

Silverman P (1978) *Animal behavior in the laboratory.* New York: Pica Press.
An introductory textbook for the experimental study of animal behavior, focusing primarily on rats. In a chapter on ethics (Chapter 4, "Some questions of ethics and experimental design," pp 25–34), the author discusses how to minimize suffering and advocates using a cost-benefit analysis in planning each experiment.

Articles

American Physiological Society (1978) Guiding principles in the care and use of animals. *Physiol 21*,3:27.
The principles state that animals must be lawfully acquired and "kindly treated." They also address the use of anesthetics, postoperative care, and student use of animals. These principles were revised in 1980 (see American Physiological Society, 1981).

Caplan AL (1978) Rights language and the ethical treatment of animals. In: McCullough LB and Morris JP III (ed) *Implications of history and ethics to medicine—Veterinary and human.* College Station, TX: Texas A&M Press, pp 126–135.
The author is a moral philosopher who rejects most criteria that have previously been suggested for possession of rights. He proposes an alter-

native criterion for moral relevance: "the fact that a being has a degree of mentality sufficient to have wants, desires, and hopes, and to engage in speculations as to how these longings might be satisfied or fulfilled." Implications for judging the morality of animal experimentation are considered in this light.

Eskridge NK (1978) Researchers concerned about changes in animal care policy. *BioSci 28*,6:409–410.
Report on the response of biomedical researchers to proposed changes in the NIH guide. Researchers, who are described as defensive, are alarmed that the guidelines may become regulations. Two bills being considered by Congress are also discussed.

Fox MW (1978) Humane ethics in veterinary education. *J Vet Med Educ* 5,2:98–101.
Examines the ethical responsibilities of veterinarians and urges that greater emphasis be placed on these responsibilities in veterinary school.

French RD (1978) Historical aspects [of animal experimentation]. In: Reich WT et al (eds) *Encyclopedia of bioethics* vol 1. New York: Free Press, pp 75–79.
Describes the history of animal experimentation and the emergence of the antivivisection movement, including a discussion of the current situation and recent developments. The author states that the antivivisection movement has had little effect on the growth of the use of animals for research, the main constraint being available resources.

Hoar RM (1978) Potential impact of GLP regs on the animal facility. *Lab Anim* (NY) *7*,7:28–30.
The author considers the Good Laboratory Practice regulations to be "a mixed blessing." Half of the article is taken up with extensive quotations from the regulations. One impact will be greatly increased costs, which could mean the end for some small academic labs.

Holden C (1978) Animal rights advocate urges new deal. *Science 201*:35.
Feature article on the work of veterinarian Michael Fox and colleagues at the new Institute for the Study of Animal Problems (ISAP) at the Humane Society of the United States (HSUS). Describes the ISAP's projects for animal welfare in research, schools, science fairs, and factory farming, all of which have given the animal rights movement "a new level of intellectual sophistication and political effectiveness."

Kitchen H (1978) On the fifth day: Animal rights and human ethics. [book review] *J Am Vet Med Assoc 173*,2:215.
Review of Morris RK, Fox, MW (eds) (1978) *On the fifth day: Animal rights and human ethics* (Washington, DC: Acropolis Books). Noting that the positions taken by many authors in this collection "will not be acceptable to all," the reviewer nevertheless recommends the book for its

"less rigid and more scholarly" approach to the subject than is found in many discussions of animal rights.

Leathers CW, Bustad LK (1978) The scope of a laboratory animal program needed at a veterinary school. *J Am Vet Med Assoc 173*,9:1218–1221.
Career opportunities for laboratory animal veterinarians are expanding, according to the authors, and programs in laboratory animal medicine need to be developed in veterinary schools. The veterinarian's ethical and legal responsibilities under the Animal Welfare Act are discussed.

Mazumdar PMH (1978) Antivivisection and medical science in Victorian society. [book review] *Bull Hist Med 52*,4:601–602.
Review of French, RD (1975) *Antivivisection and medical science in Victorian society* (Princeton, NJ: Princeton University Press). The reviewer favors the approach taken by French, which is to analyze the antivivisectionist movement in terms of sociology of science. French is quoted approvingly on his view that "'the historical interest attaching to the [antivivisectionist] movement involves not cruelty to animals, so much as tensions surrounding the roles of science and medicine in society.'"

McDonald LE, Booth NH, Lumb WV, et al (1978) Report of the AVMA panel on euthanasia. *J Am Vet Med Assoc 173*:59–72.
This is a revision of the reports published in 1963 and 1972 (see Smith CR, et al, 1972). Methods of euthanasia evaluated here that were not mentioned in earlier reports include T-61 (an injectable nonbarbiturate, nonnarcotic drug), nicotine (which is "absolutely condemned"), and microwave irradiation. Figures are included showing physiological recordings taken from four dogs undergoing euthanasia by four of the methods discussed. (The report was further updated in 1986: *J Am Vet Med Assoc 188*,3:252–268. An abbreviated version of the 1986 report is reprinted in this book as an addendum to Appendix C.)

Melby EC Jr (1978) Why laboratory animal medicine should be in a veterinary school curriculum. *J Am Vet Med Assoc 173*,9:1212–1214.
Traces the growth of the specialty in laboratory animals within the veterinary profession and the increasing responsibilities veterinarians have been given for laboratory animals under various laws and regulations. The author notes that "indeed, we now hear the words 'animal rights' being expressed both within and outside of our profession, thus bringing an entirely new set of perspectives to bear."

Merenbloom R (1978) Animal model research protocols: The importance of review. *Lab Anim* (NY) 7,2:47–49.
Emphasizes the importance of peer review of protocols: both internal, by an animal welfare committee, and external, by an expert from outside the institution. The object is to develop "a well designed and scientifically

sound proposal," but animal welfare concerns are also mentioned. The author discusses factors that animal welfare committees should consider in their reviews.

Mulder JB (1978) Review of *On the fifth day.* [book review] *Lab Anim Sci 28*:370–371.
Review of Morris RK, Fox, MW (eds) (1978) *On the fifth day: Animal rights and human ethics* (Washington, DC: Acropolis Books). Brief but highly critical review of this book, many of whose contributors question the morality of animal research and other exploitive uses of animals. The papers are described as "poorly documented" and containing "sensationally exploited statements." The reviewer, who advocates creationism, objects to the acceptance of evolutionary theory by many of the contributors and justifies killing animals by citing biblical passages.

Newton WM (1978) Environmental impact on laboratory animals. *Adv Vet Sci Comp Med 22*:1–28.
A technical article dealing with specific environmental factors (light, noise, temperature, etc) and their effects on laboratory animals. The emphasis is on obtaining valid experimental results, but humane considerations are also mentioned. A section on stress gives technical information on physiologic effects of specific stressful procedures (eg, different methods of euthanasia), which might interfere with experimental results.

Rollin BE (1978) Updating veterinary medical ethics. *J Am Vet Med Assoc 173*,8:1015–1018.
The author criticizes the American Veterinary Medical Association's "Principles of Veterinary Medical Ethics" for focusing on trivia, such as professional etiquette and advertising, instead of fundamental moral issues. Lists more than a dozen ethical questions that veterinarians ought to address, but have not. Several of these pertain to animal research.

Singer P (1978) Philosophical perspectives [on animal experimentation]. In: Reich WT et al (eds) *Encyclopedia of bioethics* vol 1. New York: Free Press, pp 79–83.
Reviews the nature and extent of animal experimentation and relevant regulations in the United Kingdom, West Germany, and the United States. The author, who considers himself a reformer rather than an absolute abolitionist, argues that alternatives could replace many animal experiments. Making sharp ethical distinctions between humans and other animals is "speciesism."

Smith DH (1978) Scientific knowledge and forbidden truths: Are there things we should not know? *Hastings Ctr Rep 8*:30–35.
This essay deals with recombinant DNA research and similar issues that raise the question of the morality of knowledge. The author refers to moral questions of animal research very briefly.

Tender costly care. (1978) *MD 22*,5:57–61.
A noncritical (unsigned) paean to the use of animals in research, emphasizing the large amount of money and effort spent on their care.

Wade N (1978) New vaccine may bring man and chimpanzee into tragic conflict. *Science 200*:1027–1030.
News feature about the need for chimpanzees in the production and testing of a vaccine against the hepatitis B virus. The dilemma is described as a conflict of interest between the needs of humans—1,500 of whom died from hepatitis B in the United States in 1976—and the needs of chimpanzees, which are already a threatened species.

1979

Articles

Caplan AL (1979) Animal ethics. [book review] *BioSci 29*,1:46.
Review of Morris RK, Fox MW (eds) (1978) *On the fifth day: Animal rights and human ethics* (Washington, DC: Acropolis Books). The reviewer judges the essays in this book to be uneven in quality, but on the whole he thinks that "this volume . . . should provide provocative reading for those willing to take the time to reflect seriously about the moral standing of those entities so tirelessly observed in the field and in the laboratory."

Cecil H (1979) Committee reports. Animal care and experimentation committee. *Physiol 22*,6:9–10.
Summarizes events at an all-day meeting held June 19, 1979, with speakers from the National Society for Medical Research (NSMR), the Scientists Center for Animal Welfare (SCAW), and the Institute for the Study of Animal Problems (ISAP).

Dewsbury D (1979) Review of *In Defense of Animals*. [book review] *Contemp Psychol 24*,10:876.
Brief review of McCoy JJ (1978) *In defense of animals* (New York: Seabury Press). Reviewer endorses the general viewpoint of the book; notes that only one chapter deals with animals used in scientific laboratories, "and that deals minimally with behavioral studies."

Henig RM (1979a) Stress science, not ethics—Animal experimentation: The battle lines soften. *BioSci 29*,3:145–148,195–196.
This lengthy news feature considers indications that scientists themselves are now "playing a leading role" in animal welfare. Such indications include the founding of the Scientists Center for Animal Welfare (SCAW), revisions in the National Institutes of Health guide, and scientific attention to developing alternatives. The author points out that many scientists

are now supporting some of the same goals that animal welfare groups have been working for—but for reasons of scientific validity.

Henig RM (1979b) Animal welfare groups press for limits on high school research. *BioSci 29*,11:651–653.
Report on the conference, "The Use of Animals in High School Biology Classes," sponsored by the Institute for the Study of Animal Problems, September 27–28, 1979, in Washington, DC. The recently revised rules for the International Science and Engineering Fair are published in a box along with the article.

Iggo A (1979) Experimental study of pain in animals: Ethical aspects. In: Bonica JJ, Liebeskind JC, Albe-Fessard DG (eds) *Advances in pain research and therapy* vol 3. Proceedings of the 2nd World Congress on Pain, Montreal, Aug 27-Sept 1, 1978. New York: Raven Press, pp 773–778.
Since we assume that animals experience pain, the author argues that this raises questions of morality or ethics in painful experiments. He presents a defense of animal experiments on deontological grounds, that is, based on our duties and moral obligations. However, pain researchers must also establish some ground rules to avoid inflicting unjustified pain.

Marcuse FL, Pear JJ (1979) Ethics in animal experimentation: Personal views. In: Keehn JD (ed) *Psychopathology in animals.* New York: Academic Press, pp 305–329.
This article constitutes the final chapter in a volume of contributions illustrating recent advances in the study of such animal psychopathologies as experimentally induced psychosis, drug addiction, epilepsy, and depression. The authors' point of view is that ethical systems develop through cultural evolution to ensure the culture's survival, and that "any type of potentially informative experiment" is justified with animals because "according significant moral rights to animals appears not to have survival value for the culture." Concern for animal welfare is not precluded, however.

Visscher MB (1979) Animal rights and alternative methods—Two new twists in the antivivisection movement. *Pharos 42*,4:11–19.
Examines how the arguments against animal experimentation have changed over the past 100 years: The new arguments are based on animal rights and alternative methods. The author rejects both of these arguments and urges biomedical researchers to speak up in defense of animal experimentation. Extensive excerpts are included from a Universal Declaration of the Rights of Animals presented to UNESCO by a group called the International Association Against Painful Experiments on Animals (IAAPEA); this is only one of the "insidious attacks on biomedical investigators" deplored in the article.

1980

Books

Dawkins MS (1980) *Animal suffering: The science of animal welfare.* New York: Methuen, Inc (also published in London by Chapman and Hall).
Deals with the problem of how to recognize animal suffering. Written for the nonspecialist, but includes full scientific references. The author, an Oxford biologist, argues that it is possible to study the subjective experiences of animals scientifically, and examines the strengths and weaknesses of various methods for assessing animal suffering. A concluding chapter provides guidelines in the form of a checklist that can be used to decide in any given case whether an animal is suffering.

Pratt D (1980) *Alternatives to pain in experiments on animals.* New York: Argus Archives.
A sequel to the author's 1976 book (qv), which is now out of print. Brings the earlier volume up to date and uses examples of painful experiments from throughout the United States. An opening chapter on pain discusses the nature of animal suffering, the psychophysiology of pain, pain-relieving drugs, and types of experiments most likely to cause suffering. Following chapters address specific types of experiments, arranged alphabetically, for which alternative methods are suggested to eliminate the need for animals or to reduce their suffering.

Roberts C (1980) *Science, animals and evolution: Reflections on some unrealized potentials of biology and medicine.* Westport CT: Greenwood Press.
Some chapters were previously published as articles in Danish or English, from 1969 to 1972. The author is highly critical of "the scientific conscience" on spiritual grounds. Four chapters deal with animal experimentation, including a reprint of a debate between the author and the neurosurgeon Robert White (Chapter 10), whose animal experiments are criticized in Chapter 9. Another chapter (Chapter 8) is an essay review of Godlovitch S, Godlovitch R, Harris J (eds) (1971) *Animals, men and morals: An enquiry into the maltreatment of non-humans* (London: Gollancz). The author takes an unusual position for a critic of animal experimentation in that she does not advocate the development of alternative methods because the latter would still promote "the increasing domination of the world by the scientific outlook."

Articles

Besch EL (1980) Environmental quality within animal facilities. *Lab Anim Sci 30*,2,pt 2:385–406.
A technical article on environmental factors such as heat, humidity, and air distribution. Discusses the impact of NIH and legislative regulations.

The author believes that these involuntary controls have had "beneficial effects," but that the best approach would be voluntary regulation established by the biomedical community itself.

Bowd AD (1980a) Ethics and animal experimentation. *Am Psychol 35*,2:224–225.
Indicts psychologists for using animals in their research without giving much attention to ethical questions. The author argues that most psychological experiments in which pain is inflicted on animals are not justified. Criticizes the American Psychological Association's code of ethics; advocates alternative experimental techniques.

Bowd AD (1980b) Ethical reservations about psychological research with animals. *Psychol Rec 30*,2:201–210.
This is an expanded version of an earlier article (see Bowd AD, 1980a), and the opening salvo in what turned into an extended and hostile exchange between Bowd and other researchers (see Gallup GG, Suarez SD, 1980a, 1980b). Bowd argues strongly that "sentient creatures have interests deserving of consideration," but psychologists have ignored this proposition. Discusses the role of language in creating and maintaining "nonscientific distinctions between human beings and animals." Suggests an alternative ethical approach.

Bowd AD (1980c) Reply to Gallup and Suarez. *Psychol Rec 30*,3:423–425.
Defends his earlier paper (see Bowd AD, 1980b) against the criticisms of Gallup and Suarez (see Gallup GG, Suarez SD, 1980a). Characterizes the point of view of his critics as "the absence of an ethic."

Bradbury RP (1980) The Good Lab Practice regulations: How to use them. *Lab Anim* (NY) *9*,4:30–33.
A spokesman for the FDA explains the background of the Good Laboratory Practice (GLP) regulations and their intent. Each item pertaining to animal care is discussed.

Brewer NR (1980) Personalities in the early history of laboratory animal science and medicine. *Lab Anim Sci 30*,4,pt.2:741–758.
Describes conditions that existed in laboratory animal care prior to the mid-1940s and gives biographical sketches of 11 men who have made contributions to the field. Devotes half a page to problems caused by antivivisectionists during that early period.

Broad WJ (1980) Legislating an end to animals in the lab. *Science 208*:575–576.
News report on the Research Modernization Act (HR 4805, introduced by Representative Fred Richmond). Describes wide public support for the bill, which is characterized as "naive and sweeping," and the strong opposition it faces from the scientific community.

Burghardt GM, Herzog HA (1980) Beyond conspecifics: Is Brer Rabbit our brother? *BioSci 30*,11:763–768.
The ethical costs of all uses of animals are considered, with most attention given to animal research. The authors, both psychologists, analyze the criteria people use to make ethical decisions about animals, and the inconsistencies and contradictions in the way people use these criteria.

Cooper JAD (1980) On the use of animals in scientific research. *J Med Educ 55*,1:72–73.
Writing as President of the Association of American Medical Colleges (AAMC), the author explains the AAMC's position opposing passage of the Research Modernization Act, which would "hinder the scientific research enterprise in this country." The most objectionable provision of the bill is the proposed diversion of 30% of research funds for the development of alternatives.

Curtis SE (1980) Who should regulate animal care and research? ARCAS's opinion. *J Anim Sci 51*,2:479–482.
The American Registry of Certified Animal Scientists (ARCAS) believes that animal scientists are the people who ought to be leading the animal welfare movement, but so far they "have been slow to join, much less to lead," this movement. The author warns that if animal scientists do not take the initiative, they will find themselves regulated by others.

Dodds JW, Abelseth MK (1980) Criteria for selecting the animal to meet the research need. *Lab Anim Sci 30*,2,pt 2:460–465.
Criteria for selection of an animal model are listed: The first one is the consideration of alternatives. The authors urge scientists not to be defensive, but to "lead the way in developing realistic guidelines and standards" for conducting humane biomedical research.

Foster JC, Meyers NM (1980) Regulatory considerations in the transportation of laboratory rodents. *Lab Anim Sci 30*,2,pt 2:312–322.
A detailed discussion of the transportation regulations of the Animal Welfare Act. The authors consider most of the regulations ill-conceived, impractical, and harmful to the laboratory animal industry. They resent the implication that the industry cannot regulate itself, and they argue that studies are needed to determine animals' transportation needs.

Fox MW (1980) Regulation of animal care and research? Animal science, welfare and animal "rights." *J Anim Sci 51*,2:483–491.
This article deals only with farm animals, including those used for research. Argues that social and emotional needs of farm animals ought to be investigated; environmental variables that affect animal behavior ought to be taken into account. Animals have the right not to be caused to suffer, or at least humans have a moral obligation not to cause them to suffer.

Gallup GG, Suarez SD (1980a) On the use of animals in psychological research. *Psychol Rec 30*,2:211–218.
Response to Bowd's article in the same issue (see Bowd AD, 1980b). The authors defend the use of animals in psychological research as a "moral imperative." Sociobiology is used to support the contention that "by virtue of our highly species-specific sense of reciprocal altruism, human morality and ethics may simply be irrelevant when applied to any species other than our own."

Gallup GG, Suarez SD (1980b) Reply to Bowd. *Psychol Rec 30*,3:427–428.
Response to Bowd's response (see Bowd AD, 1980c) to their criticism (see Gallup GG, Suarez SD, 1980a) of his original article (see Bowd AD, 1980b). The authors protest that Bowd distorted their views.

Grafton TS (1980) The founding and early history of the National Society for Medical Research. *Lab Anim Sci 30*,4,pt 2:759–764.
Outlines the efforts of the National Society for Medical Research (NSMR), founded in 1946, "to inform the public regarding the necessity, humane character and accomplishments of animal experimentation." Includes the rules for animal experimentation published in 1929 by the American Association for Medical Progress.

Hoff C (1980) Sounding board. Immoral and moral uses of animals. *N Engl J Med 302*,2:115–118.
A philosophical essay. The author believes that "animals are entitled to moral consideration" based on their ability to suffer pain. The argument is that many animal experiments cannot be justified, but some can because "human lives are generally worthier than animal lives."

Lepore PD, Crawford LM (1980) Regulation of animal care and research. The final order on Good Laboratory Practices. *J Anim Sci 51*,2:474–478.
Spokespersons for the FDA's Bureau of Veterinary Medicine discuss the deficiencies uncovered during FDA inspections that led to the development of the Good Laboratory Practice (GLP) regulations, comments and criticisms that have been made, and details of the regulations. A table is included showing the contents headings for each subpart of the regulations.

McPherson C (1980) Regulation of animal care and research? National Institutes of Health's opinion. *J Anim Sci 51*,2:492–496.
Brief summary of the NIH position, which is "based on the premise that healthy, well-cared for animals are necessary for valid research and that those who use animals have a responsibility for the animals' care and well-being." Appendices contain the NIH principles and the table of contents of the NIH guide.

Oltjen RR, Kvasnicka WG (1980) Who should regulate the care of farm animals used in research? An opinion of USDA scientists and administrators. *J Anim Sci 51*,2:497–501.

The USDA system of regulating farm animal research relies on self-regulatory mechanisms. The authors argue that this system works well; no further regulation is needed or advised.

Phifer DC (1980) Why technician training is important: The effects of the GLPs and research's modernization on the technician in industry. *Lab Anim* (NY) *9*,5:43–45.
The author points out that laboratory animal care has become highly specialized and much of the responsibility for compliance with Good Laboratory Practice regulations falls on technicians. The American Association of Laboratory Animal Science (AALAS) offers programs that provide a good way for technicians to increase their skills. (The author is an AALAS-certified laboratory animal technologist.)

Pryor WH (1980) Quality in laboratory animal care: The American Association for the Accreditation of Laboratory Animal Care (AAALAC) perspective. *Lab Anim* (NY) *9*,5:40–41.
Explains the goals of the American Association for the Accreditation of Laboratory Animal Care (AAALAC) in assuring "optimal laboratory animal care, humane treatment, protection of personnel and control of variables" through its accreditation program. Reviews the current status of this program, including the information that 400 institutions now have AAALAC accreditation. [Note: By December 31, 1987, 521 institutions had been accredited. See Chapter 1.]

Rowan AN (1980) The new federal bills: What they are and what they mean for animal research. *Lab Anim* (NY) *9*,4:25–28.
Three bills that deal with alternative methods (the Drinan bill, the Weiss bill, and the Research Modernization Act) and one amendment to the Animal Welfare Act (the Schroeder bill) are discussed. The latter is the only bill explicitly endorsed, but the concept of encouraging the development of alternatives is strongly supported.

Sherman PW, Sherman CK (1980) On a philosophical level. [book review] *BioSci 30*,5:351.
Review of Midgeley M (1978) *Beast and man* (Ithaca, NY: Cornell University Press). The reviewers think this book is "an ambitious first effort to deal with human nature in a cross-disciplinary perspective," and they praise Midgeley as "one of the first philosophers to appreciate the potentially profound implications of Darwinism for her field," but they criticize "her lack of understanding of evolutionary theory."

Townes J (1980) Federal regulations: An overview. *Lab Anim* (NY) *9*,4:16–22.
A detailed description of the provisions of the Animal Welfare Act of 1966 and subsequent (1970 and 1976) amendments, the NIH guidelines, and Good Laboratory Practice regulations. Problems of administering and enforcing the act are discussed.

Trull FL (1980) The congressional bills on alternatives to animals: How should the animal research community respond? *Lab Anim* (NY) *9*,3:19–20.
Statement of the position of the Research Animal Alliance on three bills currently before Congress. The bills, all intended to promote the development of alternatives to live animals in research, are summarized. The author believes that Congressional leaders "generally have no idea of the major ramifications and setbacks [these bills] would place on bio-medical science and the advance of basic science."

Ullrey DE (1980) Regulation of animal care and research? Viewpoint of an agricultural experiment station scientist. *J Anim Sci 51*,2:502–504.
Endorses the viewpoint that man has a "moral reponsibility" to manage the environment correctly, but disagrees with animal philosophers such as Peter Singer. The author believes that more government regulations are not needed.

Ward JW (1980) How the GLPs changed one drug company's research operations. *Lab Anim* (NY) *9*,4:35–43.
A technical, detailed report on the procedures instituted at the AH Robins Co., Richmond, VA, to ensure compliance with the Good Laboratory Practice regulations. The focus is on record keeping, not humane concerns. The author considers the changes made as a result of the GLPs to be "a very positive benefit."

Woods JE (1980) The animal enclosure—A microenvironment. *Lab Anim Sci 30*,2,pt 2:407–413.
Technical paper on the criteria for controlling the microenvironment of the laboratory animal cage; one of these criteria is the protection and comfort of the animal. Notes that current standards and guidelines are not adequate and calls for better voluntary standards.

1981

Book

Basson MD (ed) (1981) *Rights and responsibilities in modern medicine: Progress in clinical and biological research*, vol 50. New York: Alan R Liss, Inc.
Proceedings of the "Fourth Conference on Ethics, Humanism, and Medicine," held at the University of Michigan, November 10, 1979. A session entitled "Animal Rights and Animal Experimentation" featured discussion of the ethics of a hypothetical research proposal to study induced coronary thrombosis in monkeys. The session was introduced by Basson (pp 65–66); the proposal was presented by Bennett J Cohen (pp 67–68); opposing arguments were presented by Tom Regan (pp 69–83) and Cohen (pp 85–92); and the discussion was summarized by Basson (pp 93–94).

Articles

Adams PM (1981) The scientist's concern for animal welfare. *Psychopharmacol Bull 17*,2:91–93.
Discusses the growth in scientific concern for animal welfare and urges scientists to explain to the public the advances achieved through animal research. The author advocates "humane treatment of animals" in order to ensure valid experimental results, but leaves moral questions ambiguous.

American Physiological Society (1981) Guiding principles in the care and use of animals. *Physiol 23*,3:38 (reprinted Feb 1982 [25,1:35] and June 1983 [26,3:114]).
Revised (1980) version of the "Guiding Principles" published June 1978. A paragraph has been added stipulating that animal experiments must be done "only with the purpose of advancing knowledge;" references to the NIH guide have been inserted; and a clause has been added allowing an exception to be made for the requirement that anesthetics be used ("when use of anesthetics would negate the results of the experiment").

American Psychological Association. APA Committee on Animal Research and Experimentation (1981) Guidelines for the use of animals in school science behavior projects. *Am Psychol 36*,6:686.
Noting that "more and more intermediate and secondary students are participating in classroom and science fair projects," the introductory paragraph states that "such projects should not only teach research procedures but should also introduce students to ethical issues in animal care and animal research." A 10-point list of guidelines follows. This is a revision of the APA's 1972 guidelines (qv).

Blackman DE (1981) Regulating psychological experimentation with animals in the United Kingdom. *Psychopharmacol Bull 17*,2:84–88.
Efforts underway to revise the Cruelty to Animals Act of 1876 are discussed, particularly as they apply to psychological research. Suggests that psychologists can contribute to the debate because of their sensitivity to environmental and other factors that affect pain perception.

De Paolo R (1981) Editor's note. *MD 25*,10:11.
Introduces a cover article, "The rights of animals" (qv). Stresses the importance of the issue, and comments that "the people now behind the animal-rights' controversy have earned the right to be taken very seriously indeed."

Fowler SL (1981a) Editorial: Controversy and science. *Lab Anim* (NY) *10*,1:7.
Explains why an interview with antivivisectionist Henry Spira appears in this issue. The editor expects criticism for this decision, but argues that controversy should not be avoided. (For Spira interview, see Harriton, 1981.)

Fowler SL (1981b) Editor's reports: Monkey business. [editorial] *Lab Anim* (NY) *10*,7:7,13.
Reports on the arrest of Dr. Edward Taub for abuse of monkeys in his laboratory in Silver Spring, MD. Discusses the NIH investigation into the case that resulted in termination of Taub's funding, and cites weaknesses in the Animal Welfare Act. Includes photographs of the monkeys; the pictures were taken by Alex Pacheco, the animal rights activist who brought the case to the attention of the police. (For an editorial defending publication of the photographs, see Fowler, 1982b.)

Fox MW (1981) Experimental psychology, animal rights, welfare, and ethics. *Psychopharmacol Bull 17*,2:80–84.
Charges that "needless repetition and unnecessary (unjustifiable) suffering are commonplace in experimental psychology." The author criticizes the American Psychological Association's code of ethics, charges that the peer review system is inadequate, and advocates numerous reforms in designing research, housing animals, etc.

Gallistel CR (1981) Bell, Magendie and the proposals to restrict the use of animals in neurobehavioral research. *Am Psychol 36*,4:357–360.
Using the 19th century experiments of Charles Bell and François Magendie to illustrate the point, the author argues that it is impossible to know in advance which experiments will be significant. Criticizes current legislative attempts to restrict experimentation because alternatives such as computer simulation are useless in studying the relation between the nervous system and behavior. Conclusion: "It is an affront to my own ethical sensibility to hear arguments that the suffering of animals is of greater moral weight than are the advancement of human understanding and the consequent alleviation of human suffering."

Grafton TS (1981) The role of alternatives in biomedical research. *Lab Anim* (NY) *10*,1:25–30.
Emphasizes the limitations of alternatives and the futility of trying to create alternatives through legislation. The author, executive director emeritus of the National Society for Medical Research, argues that biomedical scientists do want in vitro alternatives for scientific and economic reasons, but "new technology will not be created through an act of law or administrative regulation."

Griffen RJ Jr (1981) Animals in pharmaceutical research. *Am Pharm NS21*,6:14–18.
Policy statement adopted by the American Pharmaceutical Association in March. The four-point statement affirms the necessity for animal research, supports humane treatment, encourages the development of alternatives, and opposes restrictive legislation. The statement is followed by a discussion of pain-relieving drugs, relevant legislation, and alternative methods.

Harriton L (1981) Conversation with Henry Spira: Draize test activist. *Lab Anim* (NY) *10*,1:16–22.
Interview with Henry Spira, organizer of the Coalition to Stop Draize Rabbit Blinding Tests. Spira emphasizes the pain suffered by rabbits used for the eye irritancy tests and the possibilities of finding alternatives, as well as the triviality of some of the products tested. Following the interview is a news report on moves announced by Revlon and by the Cosmetic, Toiletry and Fragrance Association (CTFA) to support research on alternatives to the Draize test. (For an editorial defending the publication of this interview, see Fowler, 1981a.)

Held JR, Milouchine V (1981) Use and protection of laboratory animals. *Anim Regul Stud 3*:273–279.
Discusses the human-animal relationship, especially in terms of our ethical and legal responsibilities toward research animals.

Holden C (1981a) Police seize primates at NIH-funded lab. *Science 214*:32–33.
Report on the Montgomery County (Maryland) police raid on the Silver Spring laboratory of Edward Taub, in which 17 monkeys were seized. Contains a sidebar announcing the Center for Alternatives to Animal Testing, established at Johns Hopkins University with a $1 million grant from the Cosmetic, Toiletry and Fragrance Association (CTFA).

Holden C (1981b) Scientist convicted for monkey neglect. *Science 214*:1218–1220.
News feature about Edward Taub's conviction on November 23 of six counts of animal cruelty; a fine of $3,015 was suspended pending appeal. Discusses precedents set by the trial, and summarizes testimony for the prosecution and for the defense. Includes photographs of Taub and of Alex Pacheco, the animal rights activist who initiated the case.

Knobil E (1981) Testimony presented before the Science, Research, and Technology Subcommittee, House of Representatives, Oct 14, 1981. *Physiol 24*,6:14–15.
The author represented the American Physiological Society and the Endocrine Society in testimony before a hearing chaired by Representative Doug Walgren to consider seven bills pertaining to animal experimentation. The societies oppose bills that would divert funding from research to the development of alternatives, and they believe that existing regulations are adequate.

Loew FM (1981) Alleviation of pain: The researcher's obligation. *Lab Anim* (NY) *10*,5:36–38.
Discusses the problem of pain in research animals and the various regulations pertaining to it. The article points out confusion caused by discrepancies in the wording of relevant sections of the Animal Welfare Act,

the NIH principles, and the NIH guide; these are reproduced along with the article. A sidebar announces the publication of new guidelines by the International Association for the Study of Pain.

Marten M (1981) The price of polio vaccine. *Lab Anim* (NY) *10*,7:20–25 (reprinted from the British animal rights magazine, *The Beast*).
Criticizes the US and British unwillingness to switch from using the Sabin polio vaccine, which requires killing monkeys, to the Salk vaccine, which can now be produced and tested using tissue cultures. "The brutal fact is that the death of 20,000 monkeys a year is a minor consideration in the complex of factors which determine immunization policy."

Migaki G (1981) Using animals to understand disease: A history. *Lab Anim* (NY) *10*,2:30–37.
Reviews the history of the use of animals to study disease, from Jenner's development of the smallpox vaccine in 1798 to the present. The author argues that new animal models must be found for diseases that have not yet been conquered.

Neil DH (1981) The genesis of the Schroeder Amendments. *Lab Anim* (NY) *10*,1:32–42.
The author was a member of the committee that drafted an amendment to the Animal Welfare Act introduced into Congress by Representative Patricia Schroeder. Explains the provisions of the bill and why the committee thinks its enactment is necessary.

Rack L (1981) Bookshelf: *The Principles of Humane Experimental Technique* and *The Use of Alternatives in Drug Research.* [book review] *Lab Anim* (NY) *10*,4:57–58.
Reviews Russell WMS, Burch RL (1959) *The principles of humane experimental technique* (London: Methuen) and Rowan AN, Stratmann CJ (eds) (1980) *The use of alternatives in drug research* (Baltimore: University Park Press). Russell and Burch are widely cited today for their early advocation of the principle of the "three Rs"—replacement, reduction, and refinement in animal experimentation. The reviewer praises the Russell and Burch book for being "ahead of its time" and also recommends the Rowan and Stratmann volume as "an interesting introduction to the current state of alternatives research."

Rowan AN (1981a) Can alternatives be integrated with animal models? *Lab Anim* (NY) *10*,2:38–45.
Argues that "we not only *can*" integrate alternatives with animal models, "but that we *must* integrate the two." Presents an optimistic assessment of prospects for alternatives to the Draize and other tests, and notes that scientists are now taking the concept of alternatives seriously.

Rowan AN (1981b) The test tube alternative. *The Sciences 21*:16–19,34.
The concept of alternative methods is defined and examples are given.

The LD–50 test and the Draize test are criticized on scientific and humane grounds. The author concludes that interest in alternatives is growing, but there is still "reluctance [on the part of] major biomedical institutions in this country to consider the possibility of alternatives."

Samuels WM (1981) Congressional hearings on the use of animals in medical research and testing. *Physiol 24*,6:13–14.
This is the first of what became a regular news column "to keep the membership informed of legislative and other issues affecting animal experimentation." Discusses testimony offered by scientists at the hearings of the Science, Research, and Technology Subcommittee, House of Representatives, chaired by Representative Doug Walgren on October 14, 1981, to consider seven bills affecting animal research. The focus of the hearings was on why the peer review system had failed in the Taub case and on alternatives.

Sechzer JA (1981) The use of animals for experimentation in the United States. *Psychopharmacol Bull 17*,2:88–91.
Examines current legislative efforts to restrict the use of animals in research in the light of events occurring in Britain at the time the Cruelty to Animals Act of 1876 was passed. The author argues that the animal protection movement has "raised the consciousness of scientists to assume responsibility for the humane treatment of animals" and has motivated them "to assure the future of animal research."

The rights of animals: Morality or practicality? (1981) *MD 25*,10:98–103,107–108.
Describes current bills in Congress to regulate animal research and develop alternatives, reviews laws in effect, and discusses the issues from the point of view of both sides. The (anonymous) author comments that animal welfare activists have had some successes lately. Includes photographs of animals undergoing experiments. (For editorial introducing this article, see De Paolo, 1981.)

Trull FL (1981) Comment: The future of animal welfare legislation under the Reagan Administration. *Lab Anim* (NY) *10*,3:21–22.
The author notes that it is too early in the new administration to tell how specific bills will fare; predicts that, since Reagan advocates decreasing legislative regulation in general, the bills may not get very far. [Note that this prediction was proved wrong when the Animal Welfare Act Amendment of 1985 and the PHS Health Research Extension Act were passed during President Reagan's second term in office.]

Visscher MB (1981) Current attempts to prevent the use of animals in medical research. *J Am Med Assoc 245*,12:1223–1224 (reprinted in *Physiol*, June 1981 [*24*,3:47–48]).
An argument opposing two bills before the 96th Congress: HR 4805, the Research Modernization Act; and HR 6847, to amend the Animal Welfare

Act. The author argues that the bills are antivivisectionist in intent, they would impede scientific progress, and they would drive up the cost of research.

Walton S (1981) Congress puts bioassay to the test. *BioSci 31*,4:287–290. Report on the NIH conference held February 18–20, 1981, in Washington, DC, to discuss bioassay methods and alternatives to the use of live animals. Current bills before Congress to promote alternatives are summarized. Conferees agreed that alternatives cannot eliminate the use of animals, but can reduce their use in some areas.

Windle DW (1981) Lab animal research for the eighties. *Lab Anim* (NY) *10*,4:15–16.
Summary of a meeting held February 10, 1981, in Milwaukee "to explore the outlook for laboratory animal science in the next decade." Speakers reviewed current regulations and controls and urged animal researchers to become more politically active.

1982

Books

Dodds JW, Orlans FB (eds) (1982) *Scientific perspectives on animal welfare.* New York: Academic Press.
Proceedings from a conference sponsored by the Scientists Center for Animal Welfare, November 1981. The conference centered around issues of accountability, and the book is organized into four sections corresponding to the responsibilities of the investigator, of the research institution, of the funding agency, and of journal editors. A final section deals with public policy and presents a summary of recommendations. (For a review, see Samuels WM, 1983.)

Articles

Adams PM (1982) Investigator responsibilities in animal experimentation. In: Dodds JW, Orlans FB (eds) (1982) *Scientific perspectives on animal welfare.* New York: Academic Press, pp 39–42.
Urges investigators to do a better job of communicating their animal welfare concerns to the public. The author makes several specific proposals for upgrading standards and educating students in doing ethical animal research.

Baker HJ, Lindsey JR, DaRif CA (1982) Responsibilities of institutions for the welfare of experimental animals. In: Dodds JW, Orlans FB (eds) (1982) *Scientific perspectives on animal welfare.* New York: Academic Press, pp 49–53.

The authors take the position that the institution, not the individual researcher, should bear the primary responsibility for assuring that animals are treated humanely. Specific recommendations are made for fulfilling this responsibility, especially in the area of review of research proposals.

Berne RM (1982) Editorial responsibilities in the care and management of animals used in research. In: Dodds JW, Orlans FB (eds) *Scientific perspectives on animal welfare.* New York: Academic Press, pp 103–105.
Suggests several specific steps journal editors can take to ensure that animals used in research are protected "from improper treatment and unnecessary pain or anxiety." Editorial review is an important "last line of defense" in reducing abuses.

Blackmore WM (1982) Animal research ethics at the University of Southern California. *Lab Anim* (NY) *11*,4:41–43,45–47.
Describes the review system for animal research recently instituted at USC. Includes a reprint of the nine-point Code of Ethics used by USC, which is identical to the code issued by the Canadian Council on Animal Care in 1980.

Brewer NR (1982) The history of euthanasia. *Lab Anim* (NY) *11*,4:17,19.
Euthanasia of animals has been discussed in the scientific literature almost entirely in relation to the history of anesthesia, which has been known at least since Paracelsus wrote about the effects of ether in the 16th century. The author points out that the pharmacodynamic action of anesthetics, which has been investigated only since the 19th century, is important in judging how humane substances are when used for euthanasia.

Bustad LK (1982) An educator's approach to euthanasia. *Lab Anim* (NY) *11*,4:37–39.
Discusses ethical, moral, and emotional aspects of euthanasia, as it applies to pets and to research animals. An outline for a course the author teaches, entitled "Reverence for Life," is included. Also included is an eight-point statement of ethics in euthanasia.

Cunningham SG, Mitchell PH (1982) The use of animals in nursing research. *Adv Nurs Sci 4*,4:72–84.
This article is an argument for increased use of animals in research by nurse scientists. The authors review the history of animal studies in nursing research, suggest why animals have been used so little, and recommend ways in which they could be used more. Ethical issues are addressed in several contexts: For example, the authors argue that animals should not be used for burn or trauma experiments because of pain. (For a rebuttal, see Crowley MA, Connors DD, 1985.)

Dodds WJ, Orlans FB (1982) Summary of recommendations on animal experimentation. In: Dodds JW, Orlans FB (eds) (1982) *Scientific perspectives on animal welfare.* New York: Academic Press, pp 123–126.
A summary of the recommendations agreed on by participants in the "First Conference on Scientific Perspectives in Animal Welfare." Twelve specific recommendations are listed.

Eyzaguirre CE (1982) Responsibilities of funding agencies for animal experimentation. In: Dodds JW, Orlans FB (eds) (1982) *Scientific perspectives on animal welfare.* New York: Academic Press, pp 69–76.
Although "the ultimate responsibility for animal welfare rests with the individual investigator," the author believes that funding agencies also play a role by reviewing applications for excessive numbers of animals or unnecessarily severe pain. The author takes the somewhat unpopular position that rats and mice, as well as cold-blooded species and even invertebrates, deserve the same moral concern as do cats and dogs.

Fowler SL (1982a) Editor's reports. [editorial] *Lab Anim* (NY) *11*,1:7.
Criticizes the way in which Dr. Edward Taub attempted to defend himself from charges that he abused animals in his research. Taub is quoted as saying, "'I suspect that there is almost no primate or animal facility in this country that you could not go through and find five violations.'" The editorial argues that statements of this sort disparage other animal researchers.

Fowler SL (1982b) Editor's reports: An explanation. [editorial] *Lab Anim* (NY) *11*,2:7.
Defends the editorial decision in the previous issue to publish an article by two antivivisectionists (see Regan T, Jamieson D, 1982) and to publish in an earlier issue some photos of the monkeys that were confiscated from Edward Taub's laboratory (see Fowler SL, 1981b). Denies readers' accusations that "we were indicating that we supported animal rights more strongly than we supported animal research."

Fox JG (1982) Institutional responsibilities: The committee's role. In: Dodds JW, Orlans FB (eds) (1982) *Scientific perspectives on animal welfare.* New York: Academic Press, pp 59–62.
Points out that the Animal Welfare Act makes the institution, not the individual researcher, the legally responsible entity in assuring humane treatment of animals. The author advocates institutional committees with a formal review process as the "best possible safeguard against inhumane animal experimentation and possible unwarranted criticism against the institution."

Fox MW (1982) Summary of workshop on editorial responsibilities. In: Dodds JW, Orlans FB (eds) (1982) *Scientific perspectives on animal welfare.* New York: Academic Press, pp 107–108.

Summarizes the recommendations that participants made in a workshop on the responsibilities of journal editors to ensure humane treatment of experimental animals. Eight specific recommendations are listed.

Golley FB (1982) Editorial responsibilities in animal research. In: Dodds JW, Orlans FB (eds) (1982) *Scientific perspectives on animal welfare.* New York: Academic Press, pp 95–100.
The author does not think that editorial review of scientific papers submitted for publication is "an effective point of control" for animal welfare and ethical issues. The animal welfare issue is considered as part of a larger social problem that involves a "self-centered, materialistic" world view.

Haddy FJ (1982) Past president's address: I think I would rather watch them make sausage. *Physiol 25*,6:466–468.
Address by the outgoing president of the American Physiological Society (APS). Reviews APS efforts during the past year to counter increased animal welfare activity and recounts the society's successes "in halting a variety of legislative initiatives that would have restricted researchers and their use of laboratory animals." The allusion in the title is to Otto von Bismark's purported advice to people with weak stomachs not to watch the making of either sausages or new laws.

Halvorson HO (1982) The responsibility of scientists in determination of public policy for the use of animals in biomedical research. In: Dodds JW, Orlans FB (eds) (1982) *Scientific perspectives on animal welfare.* New York: Academic Press, pp 109–121.
Warns against putting too much emphasis on alternatives, but also urges scientists to take seriously the public concern for animal suffering. The author suggests some ways in which the experimental use of animals could be improved.

Hanlon CR (1982) Antivivisection in new clothes. [editorial] *Bull Am Coll Surg 67*,6:1.
Denounces current legislative bills "to curtail the use of animals in research," and stresses the value of animal experimentation in surgery. The author, director of the American College of Surgeons, warns colleagues not to be taken in by bills that emphasize alternative methods; these are "deceptive features that obscure their basic anti-science and antivivisectionist thrust."

Holden C (1982) New focus on replacing animals in the lab. *Science 215*:35–38.
This is a lengthy news feature on efforts to develop alternatives to the use of animals in research. The reporter notes the influence of the animal welfare movement but comments that scientific and economic reasons for finding alternatives are more effective. On this issue, the report finds,

"scientists and animal welfare people do not appear to be much in conflict."

Keller GL (1982) Physical euthanasia methods. *Lab Anim* (NY) *11*,4:20–26.
The methods discussed are stunning, cervical dislocation, decapitation, electrocution, microwaves, exsanguination, decompression, and gas chambers. Nine specific criteria are listed—number one is "painlessness"—for evaluating the appropriateness of each method for various species. (For chemical methods, see Lumb and Moreland, 1982.)

Kerr FWL (1982) The investigator's responsibilities in research using animals. In: Dodds JW, Orlans FB (eds) (1982) *Scientific perspectives on animal welfare.* New York: Academic Press, pp 19–31.
The author assumes a consensus on the question of justification for using animals in research and so addresses other issues, primarily pain. The paper contains a detailed consideration of ethical issues raised by specific operative procedures and ethical problems in the study of pain. Acute and chronic pain studies are discussed separately, with the latter found to be ethically unacceptable or valueless in most cases.

Kraner K (1982) Funding agency responsibilities in animal experimentation. In: Dodds JW, Orlans FB (eds) (1982) *Scientific perspectives on animal welfare.* New York: Academic Press, pp 83–84.
The author is a member of an NIH study section that reviews grant applications. In his experience, study sections do address humane issues such as the use of appropriate animal model, the use of the smallest possible number of animals, and protection from unnecessary pain. Strengthening grantee institutions' local review committees is recommended, since they form "the weakest link in the chain."

Loew FM (1982a) Animal experimentation. [book review] *Bull Hist Med 56*,1:123–126.
Review of Turner J (1980) *Animals, pain and humanity in the Victorian mind: Reckoning with the beast* (Baltimore, MD: Johns Hopkins University Press) and Dalton, JC (1980) *John Call Dalton on experimental method* (New York: Arno Press). The latter is a new edition in one volume of two books written in the 19th century by an eminent experimental physiologist; the former is an account of the struggle between science and antivivisectionists in the 19th century. The reviewer considers these books indications of "a resurgence of scholarly interest in antivivisection." Both books receive generally favorable evaluations.

Loew FM (1982b) Developments in the history of the use of animals in medical research. In: Dodds JW, Orlans FB (eds) (1982) *Scientific perspectives on animal welfare.* New York: Academic Press, pp 3–6.

A brief review of the use of animals in scientific research, beginning with Galen (129–199 AD) and continuing through the development of regulations and the recent decline in the numbers of animals used.

Lumb WV, Moreland AF (1982) Chemical methods for euthanasia. *Lab Anim* (NY) *11*,4:29,33,35.
Discusses the advantages and disadvantages of various noninhalant chemical agents of euthanasia: barbiturates, T–61, chloral hydrate, and muscle relaxants. The list of criteria for evaluating different methods makes no explicit mention of painlessness or humaneness, although "speed and reliability" are important criteria. (For physical methods, see Keller, 1982.)

Malone TE (1982) Toward refinement, replacement, and reduction in the care and use of laboratory animals. In: Dodds JW, Orlans FB (eds) (1982) *Scientific perspectives on animal welfare.* New York: Academic Press, pp 7–15.
The author, Acting Director of NIH, affirms the need for animals in research and calls for "a middle but high ground that can accommodate legitimate scientific needs as well as animal welfare concerns." Describes NIH's activities since the 1950s in developing guidelines and reviewing protocols, and announces plans for NIH site visits during the coming year.

Öbrink KJ (1982) Swedish law on laboratory animals. In: Dodds JW, Orlans FB (eds) *Scientific perspectives on animal welfare.* New York: Academic Press, pp 55–58.
Describes the Swedish system for regulating animal experiments, as mandated by the (Swedish) Animal Welfare Act of 1944 and amendments of 1979. The five-level Swedish classification for experiments, based on the degree of discomfort to the animal, is presented in a table.

Orlans FB (1982) Summary of workshop on funding agency responsibilities. In: Dodds JW, Orlans FB (eds) (1982) *Scientific perspectives on animal welfare.* New York: Academic Press, pp 85–91.
A summary of workshop discussions on the responsibilities of funding agencies for animal welfare. Recommendations are given in three general areas: (1) preaward merit review, (2) site visits, and (3) accountability.

Ramsay DJ, Spinelli JS (1982) Responsibility of funding agencies: Central or local control? In: Dodds JW, Orlans FB (eds) (1982) *Scientific perspectives on animal welfare.* New York: Academic Press, pp 77–82.
The authors recommend a mixture of central and local control such as the system adopted by the University of California School of Medicine (where both authors work). They believe that the funding agencies' responsibilities should be to "lay down a framework of regulations" that local committees have the responsibility of carrying out.

Regan T, Jamieson D (1982) The case for alternatives to laboratory animal experimentation. *Lab Anim* (NY) *11*,1:21–23.
Two philosophers examine and criticize arguments for and against the use of animals in research. They conclude that "it is wrong to use animals for scientific purposes if (a) this harms the animals and (b) a conscientious effort has not been made to explore the alternatives." (For an editorial defending the decision to publish this article, see Fowler, 1982b.)

Rowsell HC (1982) Summary of workshop on investigator responsibilities. In: Dodds JW, Orlans FB (eds) (1982) *Scientific perspectives on animal welfare*. New York: Academic Press, pp 43–46.
Summarizes points agreed on by participants in a workshop on the responsibilities of researchers for the humane use of animals. The issues addressed are investigators' responsibility for preventing pain and suffering, for sensitizing colleagues to the issues, for using random source versus purpose-bred dogs, and for ensuring accountability.

Segal E (1982) Editorial. *J Exper Analysis Behav 38*:116.
Urges scientists to question their treatment of laboratory animals more than they are in the habit of doing. Expresses sympathy with the aims of animal welfare proponents, who "do not want to ban research on animals. Rather, they ask for an end to wanton and unthinking cruelty."

Shurkus MB (1982) Animal-research bill "may not be in the best interests of mankind." *Bull Am Coll Surg 67*,6:20–22.
News update on the Walgren bill, being circulated in draft form to replace a previous bill. Most of this article consists of quotations from Dr. C Rollins Hanlon, Director of the American College of Surgeons, and others who oppose the bill.

Simmonds RC (1982) Summary of workshop on institutional responsibilities. In: Dodds JW, Orlans FB (eds) (1982) *Scientific perspectives on animal welfare*. New York: Academic Press, pp 63–65.
Summary of the views of participants in a workshop on the responsibility of the institution for humane care of animals used in research. It was agreed that the institution has the "ultimate and legal responsibility" and that an institutional Animal Care and Use Committee could be an effective mechanism for fulfilling this responsibility.

Stickrod G, Chadwick AB (1982) Methods: Pain relief for rabbits and technicians: Using Innovar-Vet to minimize blood withdrawal stress. *Lab Anim 11*,4:48–49.
This is one of the few articles found that did not discuss research animal welfare issues in the abstract, but rather provided practical advice for making a particular procedure more humane. Because "collecting blood from rabbits can be painful, from the viewpoint of both the rabbit and the technician," the authors tested the effects of a drug (Innovar-Vet) that

acts as a sedative, an analgesic, and a vasodilator. It worked so well that the rabbits did not have to be restrained and showed "no signs of pain or distress."

Thomas GJ (1982) Editorial policy for animal experimentation. In: Dodds JW, Orlans FB (eds) (1982) *Scientific perspectives on animal welfare.* New York: Academic Press, pp 101–102.
Citing his own experiences as editor of a scientific research journal, the author concludes that editors can have very little impact in enforcing standards for humane treatment of experimental animals. The most editors can do is to "codify knowledge of procedures that are already well understood by their scientific community."

Visscher MB (1982) Review of *Animal rights and human morality.* [book review] *N Engl J Med 306*,2:1303–1304.
Review of Rollin BE (1981) *Animal rights and human morality* (Buffalo: Prometheus Books). The reviewer is extremely critical of the book, calling it "pseudointellectual," "highly sentimental about all animals," and containing "logic [that] is frequently very tortuous." The reviewer states that this book's "appearance at this time is evidence that the anti-science group in America is becoming more vociferous."

Weksler ME (1982) Animal experimentation and the scientist. In: Dodds JW, Orlans FB (eds) (1982) *Scientific perspectives on animal welfare.* New York: Academic Press, pp 33–38.
Offers an example of how research on the effects of undernutrition on life span was carried out in a humane way. However, the author is not optimistic about alternatives and is "frightened by the prospect of legislation which would hamper animal research and so restrict the progress of science."

1983

Books

Kitchell RL, Erickson HH, Carstens E, Davis LE (eds) (1983) *Animal pain: Perception and alleviation.* Bethesda, MD: American Physiological Society.
The papers collected in this volume were first presented at a symposium held at the 1982 meeting of the Federation of the American Societies for Experimental Biology (FASEB). The editors note that "this is the first symposium known to us that has specifically addressed pain in animals." Several papers focus on dogs, cats, nonhuman primates, and horses. The book should be useful to scientific investigators as a source of basic information about animal pain.

Sechzer JA (ed) (1983) *The role of animals in biomedical research.* Annals of the New York Academy of Sciences, vol 406. New York: The New York Academy of Sciences.
Proceedings from a workshop on animal research held in April 1982. The 22 papers cover such topics as ethical issues, prospects for alternatives to the use of live animals in research and testing, pain in laboratory animals, animal models for specific fields, and legislative and regulatory considerations.

Articles

American Physiological Society (1983a) APS statement on NIH *Guide* revisions. *Physiol 26*,3:113–114.
The statement presented by Helene Cecil, representing the American Physiological Society (APS), at a public meeting held May 17, 1983, at the National Academy of Sciences on the preparation of the sixth edition of the NIH *Guide for the care and use of laboratory animals.* Emphasizes the necessity for animal experimentation and the APS's longstanding commitment to guidelines.

American Physiological Society (1983b) APS statements on NIH *Guide* revisions. *Physiol 26*,5:273–274.
The statements presented on behalf of the American Physiological Society (APS) by David J. Ramsay at a public meeting in San Francisco, July 11, 1983; and by Walter C. Randall at a public meeting in Rosemont, IL, July 17, 1983; to consider the proposed revisions to the NIH guide. Ramsay emphasized the APS's preference for local control over animal experimentation, and Randall emphasized the need for animals in research.

Beer C (1983) Motives and metaphors in considerations of animal nature. In: Pfaff DW (ed) *Ethical questions in brain and behavior: Problems and opportunities.* New York: Springer-Verlag, pp 125–140.
Considers the philosophical problem of other minds in reference to animals and their use in research. The author asks whether we use the logic of metaphor or the logic of motive (intentional logic) when we talk about animals or when we treat them as though they have intentions. Briefly touches on the ethical consequences of conceiving of animals as intentional agents. (For a review, see Bird SJ, 1984.)

Berman D (1983) Proceedings summary. In: Sechzer JA (ed) *The role of animals in biomedical research.* Annals of the New York Academy of Sciences, vol 406. New York: The New York Academy of Sciences, pp 223–227.
Provides a summary of the goals and objectives of the workshop. In addition to an overview of the scientific method and the need for replication, participants spoke of the contributions of animals in their disciplines, concerns of their respective groups about how the animals were

being used, and efforts underway in the various scientific specialties to minimize pain and distress to the animals without compromising scientific validity. A recurring theme throughout the meeting was a reminder that the morality of animal experimentation requires that investigators provide compelling justification for the use of animals in each (every) context.

Caplan AL (1983) Beastly conduct: ethical issues in animal experimentation. In: Sechzer JA (ed) *The role of animals in biomedical research.* Annals of the New York Academy of Sciences, vol 406. New York: The New York Academy of Sciences, pp 159–169.
Focuses on an analysis of the moral value of animals, which the author (a philosopher) sees as the crucial issue in the debate over animal experimentation. Argues that purposiveness is sufficient for conferring moral worth, but there is sometimes justification for overriding an animal's rights. Concludes that "the burden of proof always rests upon the experimenter to justify the use of animals" and that "such activity is always morally tragic."

Chambers KT, Hines C (1983) Recent developments concerning the use of animals in medical research. *J Leg Med 4*,1:109–129.
Two lawyers (law students at the time of writing) discuss the concerns of animal rights groups and scientists in an effort to meet "the legitimate concerns of each." Federal and state laws are found wanting in their protection of animals. Includes an extensive review of legal decisions showing that there is little support in the courts for granting animals legal standing or allowing humans to sue on their behalf. The authors contend that a solution is needed that will "reach behind laboratory doors" but that will not "ignore the rights of medical researchers." (See Overcast TD, Sales BD 1985.)

Cooper JAD (1983) Animals in biomedical and behavioral research. [editorial] *J Med Educ 58*,1:71–72.
Expresses the view of the Association of American Medical Colleges (AAMC) opposing the Humane Care and Development of Substitutes for Animals in Research Act introduced in the House by Representative Doug Walgren. The author suggests that greater self-regulation is a better approach, since the existing laws and guidelines are adequate.

Cooper T, Galligan JE (1983) Leaving the island behind: An approach to the animal experimentation issue. *Physiol 26*,6(suppl):359–360.
Urges scientists to become more politically active to counter the effective political activism of animal welfare people. So far, the authors charge, scientists have not responded well to attacks on animal research; they act as if they were on an island, insulated from the public.

Cooper T, Stucki JC (1983) Commentary on "Animal Research—For and against: A philosophical, social, and historical perspective." *Perspec Biol & Med 27*,1:18–21.

This article responds to an article by Rowan and Rollin (qv) in the same issue. The question, according to the authors, is not whether animal experimentation should be done, but under what conditions it should be done. They argue that more animal research is needed. A better dialogue between sides and "a more balanced presentation of the issues to the public" are foreseen, but not the end of animal experimentation.

Curran WJ (1983) Biomedical laboratories and criminal liability for cruelty to animals. *N Engl J Med 309*,25:1564–1565.
Commentary on the August 10, 1983, ruling by the Maryland Court of Appeals dismissing all charges against Edward Taub. International efforts to regulate animal research are summarized as a background to the Taub case. The author applauds the court's ruling, but thinks it should be extended to prevent other states from drafting laws that would apply to research institutions.

Dubner R (1983) Pain research in animals. In: Sechzer JA (ed) *The role of animals in biomedical research.* Annals of the New York Academy of Sciences, vol 406. New York: The New York Academy of Sciences, pp 128–132.
Discusses the ethical dilemma in using animals to study pain: how such experiments can be justified and what guidelines should be followed in carrying them out. The author recommends allowing the animal to control the level of pain received and not exposing animals to pain greater than humans would accept.

Ewald BH, Gregg DA (1983) Animal research for animals. In: Sechzer JA (ed) *The role of animals in biomedical research.* Annals of the New York Academy of Sciences, vol 406. New York: The New York Academy of Sciences, pp 48–58.
Describes how experimental research using animals has benefited the animals themselves through improved veterinary medicine and techniques. Specific benefits are discussed in cancer, hereditary diseases, reproduction, infectious diseases, parasitology, and clinical techniques.

Flowers FH (1983) Research animal care in Canada: Its control and regulation. In: Sechzer JA (ed) *The role of animals in biomedical research.* Annals of the New York Academy of Sciences, vol 406. New York: The New York Academy of Sciences, pp 144–149.
The Canadian approach to regulation of the use of animals in research and testing, based on voluntary peer review, is described as working well. The Canadian Council on Animal Care (CCAC), an independent body, oversees a system of local animal care committees that review each protocol.

Friess SL (1983) Critical issues facing animal scientists. *J Anim Sci 56*,1:217–221.

Of the nine "critical issues" discussed here, most involve technical- scientific problems; however, one is the problem posed by antivivisectionists. The author believes that pressure from these groups "must be resisted publicly." Good Laboratory Practice regulations (GLPs) are another problem issue: It is argued that they are so complex that they stifle initiative and drive up costs.

Gardner KD Jr (1983) Looking upon the water: Graphic illustrations. [editorial] *Am J Kidney Dis 3,2*:87–89.
A highly facetious discussion of a bill that would restrict animal research (the Walgren bill in the House and the Dole bill in the Senate). The author opposes the bill, a point he makes by stating sarcastically that he is in favor of it because it "means we will have to hit, kick, knife, shoot, bomb, irradiate, poison, slander and exploit our animals without conscience!"

Griffin A, Sechzer JA (1983) Mandatory versus voluntary regulation of biomedical research. In: Sechzer JA (ed) *The role of animals in biomedical research.* Annals of the New York Academy of Sciences, vol 406. New York: The New York Academy of Sciences, pp 187–200.
Regulations and legislation currently in effect are reviewed, and proposed bills introduced in the 97th Congress are discussed. The bills incorporate various combinations of mandatory and voluntary regulation, for which the authors suggest a classification system. An epilogue updates the bills' progress through May 1983.

Grodsky PB (1983) Public opinion on animal-based research: The unknown factor in ethical and policy decisions. In: Sechzer JA (ed) *The role of animals in biomedical research.* Annals of the New York Academy of Sciences, vol 406. New York: The New York Academy of Sciences, pp 157–158.
The author, a social scientist, points out the need for information about the opinions of the general public and of scientists on the ethics of animal research.

Hodos W (1983) Animal welfare considerations in neuroscience research. In: Sechzer JA (ed) *The role of animals in biomedical research.* Annals of the New York Academy of Sciences, vol 406. New York: The New York Academy of Sciences, pp 119–127.
Discusses trends in neuroscience research toward alternative methods, but emphasizes that intact animals are still necessary. The author advises researchers to justify experiments using a cost-benefit analysis and suggests specific areas in which improvements could be made. He concludes that "if we can get our colleagues and students to inject some animal welfare considerations into the process of deciding how to do or whether to do a particular experiment, we will have taken a major step forward."

Matthysse S (1983) Making animal models relevant to psychiatry. In: Sechzer JA (ed) *The role of animals in biomedical research.* Annals of

the New York Academy of Sciences, vol 406. New York: The New York Academy of Sciences, pp 133–139.
Questions the focus of much current research on new drugs to treat schizophrenia because they represent little advance over drugs already available; therefore, this research is ethically questionable as far as the animals and the patients are concerned. Suggests ways to use animals better in psychiatric drug research, making such research "more respectful of animal life."

Miller NE (1983) Understanding the use of animals in behavioral research: Some critical issues. In: Sechzer JA (ed) *The role of animals in biomedical research.* Annals of the New York Academy of Sciences, vol 406. New York: The New York Academy of Sciences, pp 113–118.
Using the example of the accidental discovery of chlorpromazine (a drug used to treat schizophrenia), the author argues that a scientist should not be asked to justify animal experiments by stating exactly what the benefits will be. The author acknowledges that animals do suffer during some experiments, but his "own moral judgement is squarely on the side of reducing human misery."

Paton WD (1983) Animal experiment: British and European legislation and practice. In: Sechzer JA (ed) *The role of animals in biomedical research.* Annals of the New York Academy of Sciences, vol 406. New York: The New York Academy of Sciences, pp 201–214.
Discusses the 1876 British law regarding animal experimentation and the issues involved in current attempts to amend it, and the Convention of the Council of Europe. The paper includes British animal-use statistics presented in four figures and two tables, plus a full-page graph showing how medicine has advanced with the growth of animal experimentation.

Randall WC (1983a) Is medical research in jeopardy? *Physiol 26*,2:73–77 (reprinted from *Hosp Prac 18*,2[Feb, 1983]).
The author discusses two threats to medical research: decreased funding and restrictions on the use of animals in research. Most of the article is devoted to the latter. The spirited defense of animal research focuses on the use of dogs in experimental heart surgery. The author himself benefited from this research when he recently underwent quadruple aortocoronary bypass surgery.

Randall WC (1983b) APS Testimony on S.657: "Improved standards for Laboratory Animals Act." *Physiol 26*,5:271–272.
The American Physiological Society's position is that "restrictive legislation is not needed," but if the Congress honestly believes that further legislation is needed, then the society gives its "qualified support" to S.657. However, the society would not support any legislation that would interfere with research methodologies and procedures.

Randall WC (1983c) Past president's address: Crises in physiological research. *Physiol 26*,6(suppl):351–356.
The author views proposed animal rights legislation as a serious threat to physiology. The dependence of medical progress on animal research is illustrated by tracing the animal experiments that eventually led to the development of the coronary bypass operation.

Ristau CA (1983) Language, cognition, and awareness in animals? In: Sechzer JA (ed) *The role of animals in biomedical research.* Annals of the New York Academy of Sciences, vol 406. New York: The New York Academy of Sciences, pp 170–186.
Starting from the premise that if "animals have awareness, cognition, or language, there is an ethical problem with the use of such animals for research," the author examines the evidence for these attributes in animals. She finds the evidence inconclusive, but thinks we must assume mental continuity between ourselves and other animals. This does not mean that animal research should not be done, but that it should be done "with restraint."

Rowan AN, Rollin BE (1983) Animal research—For and against: A philosophical, social, and historical perspective. *Perspec Biol & Med 27*,1:1–17.
Reviews the history of experimental animal research and antivivisectionism, focusing on the United States. Reasons for the present high level of public concern are discussed. The authors (a biochemist and a philosopher) conclude that the goals of animal welfare activists and scientists are no longer very far apart, so "the prospects for improving the lot of laboratory animals look much brighter." (For opposing comments, see Cooper and Stucki, 1983.)

Samuels WM (1983) Review of *Ethics and animals* and *Scientific perspectives on animal welfare.* [book review] *N Engl J Med 309*,15,pt 2:932.
Review of Miller HB, Williams WH (eds) (1983) *Ethics and animals* (Clifton, NJ: Humana Press) and Dodds JW, Orlans FB (eds) (1982) *Scientific perspectives on animal welfare* (New York: Academic Press). Most of the review is devoted to the Miller and Williams book, which is criticized for having an animal rights bias. The Dodds and Orlans book (qv) is reviewed more favorably.

Sechzer JA (1983) The ethical dilemma of some classical animal experiments. In: Sechzer JA (ed) *The role of animals in biomedical research.* Annals of the New York Academy of Sciences, vol 406. New York: The New York Academy of Sciences, pp 5–12.
An examination of the ethics of two 19th century animal experiments, both done before the discovery of anesthesia: (1) François Magendie's experiments on the nerve roots of the spinal cord, and (2) Claude Bernard's experiments with curare. The author concludes that both were

justified because of their benefits to humans, but such painful ethical dilemmas are still the problem for researchers who cannot know how significant their findings may be.

Simpson LL (1983) Animals and animal tissues in the discipline of pharmacology. In: Sechzer JA (ed) *The role of animals in biomedical research.* Annals of the New York Academy of Sciences, vol 406. New York: The New York Academy of Sciences, pp 74–81.
Argues that the use of animals in the development and testing of drugs is essential and that it would be unethical to eliminate animal research in pharmacology. Nevertheless, the author concedes that "the discipline of ethics can contribute . . . to an understanding of acceptable ways to do animal research."

Stark DM, Shopsis C (1983) Developing alternative assay systems for toxicity testing. In: Sechzer JA (ed) *The role of animals in biomedical research.* Annals of the New York Academy of Sciences, vol 406. New York: The New York Academy of Sciences, pp 92–103.
The authors present an optimistic assessment of the possibilities for the development of in vitro tests to replace some tests currently done in vivo in toxicology testing. Examples of several such alternatives developed recently are described, including the authors' own efforts to replace the Draize irritancy assay with in vitro cellular systems.

1984

Books

Rowan AN (1984a) *Of mice, models, & men: A critical evaluation of animal research.* Albany, NY: State University of New York Press.
An introduction to the issues surrounding animal experimentation and the historical background of the controversy; focuses on England and the United States. A major section of the book is devoted to alternatives, with an optimistic assessment of the prospects for developing alternatives to live animals in many areas of research. The author argues that ethical issues should be addressed more coherently than they are at present and concludes with specific recommendations for change. This is the best single source for information and rational discussion of the entire range of problems presented by the research animal welfare debate. (For a review, see Mulder JB, 1984.)

US Dept of Health and Human Services, Public Health Service, National Institutes of Health, National Library of Medicine. (1984) *Laboratory animal welfare.* Specialized Bibliography Series. Bethesda, MD: Reference Services Division, National Library of Medicine.
A bibliography on laboratory animal welfare compiled by Fritz P Gluckstein, DVM, Coordinator for Veterinary Affairs, USDHHS. Contains 85

selected annotated citations: books and monographs published between 1975 and October 1983; and journal articles published between 1977 and September 1983.

Articles

Adams PM (1984) Do animals have rights? [book review] *Contemp Psychol 29*,5:401.
Review of Miller HB, Williams WH (eds) (1983) *Ethics and animals* (Clifton, NJ: Humana Press). The reviewer is critical of some of the papers included, but concludes that "this work should be a valuable contribution to the literature on ethics as it relates to the treatment of animals."

American Physiological Society (1984a) Proposed change in APS guiding principles in the care and use of animals. *Physiol 27*,3:141.
Members are requested to comment on a proposed revision of the APS Guiding Principles in the Care and Use of Animals regarding the use of paralytic drugs. The revision specifies that the animal must already be unconscious from a general anesthetic before a paralytic drug is administered, and the paralysis must wear off before the animal regains consciousness, except under certain conditions.

American Physiological Society (1984b) APS position on the actions against Dr. Edward Taub. *Physiol 27*,3:144.
In response to Taub's request that the APS support his position, the APS formed a committee that spent 3 months investigating the case. The committee's four-point conclusion generally offers Taub the APS's strong support, but leaves open the question of whether there were deficiencies in the animal facilities where he worked.

Barger AC (1984) Some of Carl Walter's contributions to Harvard Medical School and Hospitals. *Am J Surg 148*,5:578–580.
A testimonial to Dr. Walter, whose "world famous sterile dog surgery course" was attended by the author when he was a Harvard student. Walter believed that learning surgical techniques on dogs "elicit[s] the urge to care" in fledgling surgeons. Much of the paper praises Walter's activities in defense of animal experimentation.

Bird SJ (1984) Ethics in neuroscience. [book review] *Science 224*:595–596.
Review of Pfaff DW (ed) (1983) *Ethical questions in brain and behavior* (New York: Springer-Verlag). Concerning the only chapter in the volume that deals with the ethical issues of animal research, the reviewer writes that "Colin Beer provides a fascinating exploration of intentionality but skirts the ethical issues that this topic raises." (See Beer C, 1983.)

Cohen BJ (1984) Public affairs: Pound animals for research and education. *Physiol 27*,4:190–191 .

Gives some historical background on the use of pound animals for scientific research and the public opposition to it. The author argues for the continued use of pound animals for research on the grounds that (1) obtaining dogs from other sources is too expensive, (2) prohibition of the use of pound animals is just a first step toward prohibition of all animal research, and (3) pound animals are to be killed anyway.

Coile DC, Miller NE (1984) How radical animal activists try to mislead humane people. *Am Psychol 39*,6:700–701.
A defense of psychological animal research against charges published by an animal rights group, Mobilization for Animals. The authors conducted a survey of every animal experiment reported in American Psychological Association (APA) research journals over 5 years (1979–1983) and did not find one example of any of the alleged abuses. They conclude that, if abuses did occur, "they certainly were infrequent, and it is extremely misleading to imply that they were typical of experimental psychology."

Erickson HH, Kitchell RL (1984) Pain perception and alleviation in animals. *Fed Proc 43*,5:1307–1312.
Reviews advances in scientific knowledge of the mechanisms of pain over the past 2 decades. The authors explain that nociceptors or pain receptors are widespread in the skin and tissues of animals, and chemical mediation of nociceptor excitation may provide a key to understanding the peripheral phenomena related to pain. As these mechanisms are better understood, alleviation of pain in experimental animals can be placed on a much firmer scientific basis.

Fishman AP (1984) Past president's address: It was a good year. *Physiol 27*,6:405–412.
About a fourth of this address is devoted to the animal welfare controversy. This section focuses on terrorist activities conducted by antivivisectionist groups and includes a table of all raids on laboratories carried out between 1981 and June 1984. Describes efforts of the American Physiological Society (APS) to counter these activities.

Flecknell PA (1984) The relief of pain in laboratory animals. *Lab Anim* (NY) *18*,2:147–160.
A review of various types of analgesic drugs available for animals and the experimental evidence for their efficacy in laboratory species. Recommends buprenorphine as the most generally useful analgesic agent at present because it is effective and one of the few long-lasting analgesics in a wide range of species.

Fox JL (1984a) Lab animal welfare issue gathers momentum. *Science 223*:468–469.
Summarizes recent legislative developments on national and state levels, and concludes that animal welfare groups have been successful lately in

getting serious consideration for legislation pertaining to the use of animals in research. Reviews responses from the scientific community.

Fox JL (1984b) Changes in animal care policy proposed. *Science 224*:364–365.
Proposed revisions in the NIH guidelines are described: The major change is the new requirement that institutions establish "committees with lay members to review individual grant proposals." The proposed changes face criticism from research institutions and animal welfare activists.

Fox JL (1984c) Lab break-in stirs animal welfare debate. *Science 224*:1319–1320.
Report on the break-in by the Animal Liberation Front (ALF) at the University of Pennsylvania Medical School. Thirty-three videotapes of head trauma experiments performed on baboons were stolen. The NIH will investigate the ALF's claim that the experiments violated animal welfare procedures.

Fox JL (1984d) Animal rights bill defeated in California. *Science 224*:1414.
This news report covers the following events: the defeat of a California bill that would have prohibited the use of pound animals for research, the defeat of a Maryland bill that would have extended the state's anti-cruelty statute to research laboratories, and the decision of HHS to uphold NIH's decision to withdraw funding for Edward Taub.

Fox JL (1984e) Rifkin takes aim at USDA animal research. *Science 226*:321.
Jeremy Rifkin has filed a suit against the USDA's gene-transfer program in which growth hormone genes would be transferred from other species into sheep and pigs. Rifkin and Michael Fox (of the Humane Society of the United States), who has joined him in the suit, object to "crossing of species barriers in mammals" on environmental, economic, and ethical grounds.

Goldberg AM (1984) The Johns Hopkins Center for Alternatives to Animal Testing. *Fund Appl Toxicol 4*,4:517–518.
Describes the center's efforts to develop new methods that do not use whole animals for testing the toxicity of chemicals, consumer products, and drugs. The center's goal is to provide superior methods that will replace existing tests.

Hannah HW (1984) Legal brief: Animal trials. *J Am Vet Med Assoc 185*,9:955.
An account of the medieval trials of animals that took place in France, Germany, and Italy (summary taken from Evans EP [1906] *The criminal prosecution and capital punishment of animals*, New York: E.P. Dutton). "The prime purpose of this article is to remind modern man that his predecessors of 600 years ago had some notions about animal rights (they

afforded them trials) and also about animal duties (they hanged them for what they concluded were their transgressions)."

Held JR, Doherty J, King NB (1984) Council report: Societal benefits derived from animal research. *J Am Vet Med Assoc 185*,3:268–269.
A brief review of some of the ethical issues of animal research, followed by an argument for the indispensability of animals in research and a recital of some of the major medical advances made possible through animal experimentation.

Hess CE (1984) Freedom of enquiry: An endangered species. [editorial] *Science 225*:273.
Efforts by animal welfare groups to restrict the use of animals in scientific research are among several "antiscience activities" that threaten the freedom of enquiry, according to Hess. Urges readers to "invest our energies on all fronts" to oppose these activities.

Jacobs FS (1984) A perspective on animal rights and domestic animals. *J Am Vet Med Assoc 184*,11:1344–1345.
Disagrees with animal rights activists; defends confinement farming, the use of animals for research and teaching, and cropping dogs' ears. The author argues that "domestic animals exist in this world because they fulfill man's needs . . . [and] they would not exist if man did not exist."

Keenan KN (1984) From *Air Force Times* (9 April 1984). [reprint] *Physiol 27*,5:350.
The author, a USMC captain and a graduate of the Uniformed Services University of the Health Sciences School of Medicine, defends the use of live animals in military medical training in the treatment of high-velocity wounds.

Key JD, Rodin AE (1984) William Osler and Arthur Conan Doyle versus the antivivisectionists: Some lessons from history for today. *Mayo Clin Proc 59*,3:189–196.
A historical vignette of 18th and 19th century experimental science and its antivivisectionist opponents, focusing on England. A major section of the article describes the efforts of Osler and Doyle on behalf of animal research—"and both were moral, sensitive, and humane physicians."

Moss TH (1984) The modern politics of laboratory animal use. *Sci Tech Human Values 9*,2:51–56 (reprinted in *BioSci 34*,10:621–625).
Using events at a recent conference ("Standards for Research with Animals: Current Issues and Proposed Legislation," sponsored by Public Responsibility in Medicine and Research) as evidence, the author suggests that the long-standing conflict between animal welfare advocates and the scientific community may now be on the verge of a resolution. "That optimism is based, in part, on a crucial shift of scientific community opinion. . . . The shift involves abandonment of the point of view that

concern for laboratory animal welfare is a minor or eccentric sideshow in the overall progress of biomedical research."

Mulder JB (1984) Book review: *Of mice, models and men: A critical evaluation of animal research.* [book review] *J Am Vet Med Assoc 184*,12:1466.
Review of Rowan AN (1984) *Of mice, models and men: A critical evaluation of animal research* (Albany, NY: State University of NY Press). The reviewer concludes this highly critical appraisal with the comment that "many readers, especially veterinarians and other scientists, will not agree with the tenor of this book." The reviewer does not agree with it either, because "the content overwhelmingly decries the use of animals in teaching and research programs and condemns much current research methodology." (See Rowan AN, 1984a.)

Myers D (1984) Animal rights, academic freedom, and progress in biomedical research. [editorial] *Cancer Investig 2*,6:497–498.
Emphasizes the importance of taking the animal rights movement seriously and the need to educate the public about the essential role that animals play in biomedical research. "If we as scientists are to succeed, an unremitting educational effort on our part will be required."

Purchase IF (1984) The future of animals in research and teaching. The European scene. *Fund Appl Toxicol 4*,4:519–530.
Reviews opinions and legislation in European countries requiring safety tests of chemicals and regulating the use of animals. Discusses the Convention on the Use of Vertebrate Animals for Experimental Purposes about to be ratified by the Council of Europe. The main issues are discussed, and the example of acute toxicity testing is considered.

Ritvo H (1984) Plus ça change: Antivivisection then and now. *Sci Tech Human Values 9*,2:57–66 (reprinted in *BioSci 34*,10:626–633).
Discussion of a conference entitled "Standards for Research with Animals: Current Issues and Proposed Legislation," which was sponsored by Public Responsibility in Medicine and Research. The author believes that scientists underestimate the moral commitment of animal welfare advocates and seeks to show this by describing the historical roots of the antivivisectionist movement in Victorian England, especially in the movement's opposition to Pasteur's work on rabies.

Rosner F (1984) Animal experimentation. The Jewish view. [editorial] *Arch Intern Med 144*,5:927–928.
Discusses how Jewish law regards the question of animal experimentation. Animal experiments are sanctioned only if they are "designed to yield information that might lead to cure of disease" and only if pain is minimized, but "not to simply satisfy intellectual curiosity."

Rowan AN (1984b) The future of animals in research and training. The search for alternatives. *Fund Appl Toxicol 4*,4:508–516.

Animal welfare protests have focused on toxicology testing in pressing for the development of alternative methods. The prospects for alternatives in carcinogen detection, Draize tests, and LD–50 tests are discussed.

Samuels WM (1984a) Public affairs: Congressional amendment proposed by APS would punish lab intruders. *Physiol 27*,5:351.
A news column that includes the testimony presented by WR Randall on behalf of the American Physiological Society (APS) at a public hearing on HR 5725, the Improved Standards for Laboratory Animals Act. In his testimony opposing the bill, Randall presented a proposed Amendment to the Animal Welfare Act, previously approved by the APS, which would provide penalties for breaking into federally funded research facilities.

Samuels WM (1984b) Public Affairs: Reporters and researchers both to blame in telling the real story. *Physiol 27*,6:403–404.
This column is comprised entirely of a speech by Walter C Randall. Randall attacks the media for "completely irresponsible reporting of the news" about animal experimentation, and he chides fellow scientists for their complacency.

Science and the citizen: Man or mouse? [unsigned editorial] (1984) *Sci Am 250*,5:62,66.
Outlines the medical benefits of animal experimentation and the current efforts to restrict it. Describes the work of the Foundation for Biomedical Research, which has been organized to respond to these threats.

Swan KG, Swan RC (1984) Animal research—Wound ballistics: Man's best friend or man? *Physiol 27*,5:347–350.
The authors argue that the use of dogs for research and training in wound ballistics is necessary. They are concerned about attempts to prohibit the use of live animals for this purpose or to substitute other animals for dogs. The problems associated with high-velocity wounds are discussed through a detailed explanation of the ballistic phenomena involved.

Trull FL, Kalikow NA (1984) Animal rights movement: A threat to biomedical research? *Cancer Investig 2*,6:479–482.
A warning to researchers that the animal rights movement is sophisticated, well organized, and well financed. The authors conclude that "unless and until researchers themselves recognize the animal rights movement as a potential threat to their ability to use animal models in research, further restrictions on biomedical research will be implemented."

Woodhouse RJ (1984) Animals in medical research: Controversy in Europe and the United States. *NY State J Med 84*,11:574–577.
Mainly concerned with events in England, this article includes a section headed "Reactions from the United States." The author sees the central question as "whether the amelioration of one evil (disease) is sufficient

excuse for the perpetration of another (vivisection)." Explores many specific issues in detail.

Young C (1984) Imperatives in research animal use—Scientific needs and animal welfare. *J Appl Physiol Respir Environ Exercise Physiol 57*,6:1923–1924.
Report written by a representative of NIH, Office of Protection from Research Risks (OPRR), summarizing the NIH-sponsored "National Symposium on Imperatives in Research Animal Use: Scientific Needs and Animal Welfare," held April 11–12, 1984, at the National Academy of Sciences. A concluding speaker mentioned, "the conflict between the needs of science and the welfare of animals, which was a recurring theme throughout the symposium."

Zola JC, Sechzer JA, Sieber JE, Griffin A (1984) Animal experimentation: Issues for the 1980s. *Sci Tech Human Values 9*,2:40–50.
The authors review the historical background of animal experimentation, the present controversy, and US legislation. The major portion of the article discusses underlying values, analyzes legislative and regulatory alternatives, and suggests that the best way to resolve the conflict is to achieve a "successful balance of scientific and humanitarian values," preferably through "effective self-regulation by professionals."

1985

Book

Moberg GP (ed) (1985) *Animal stress.* Bethesda, MD: American Physiological Society.
This book grew out of a multidisciplinary symposium held July 1983 at the University of California at Davis to discuss problems involved in defining and measuring stress in animals. The papers are arranged in four sections: "Determining Animal Well-Being," "Stress in Animals," "Effects of Stress on Well-Being," and "Well-Being of Laboratory Animals." The latter consists of two papers that emphasize the importance of this topic to the development of regulations and guidelines affecting animal research. (See Besch EL, 1985; Irving GW III, 1985; Kitchell RL, Johnson RD, 1985.)

Articles

American Veterinary Medical Association Animal Welfare Committee (1985) Research animal statement proposed. *J Am Vet Med Assoc 187*,10:991–992.
Proposed revision of the American Veterinary Medical Association (AVMA) position (adopted in 1982) on the use and regulation of research

animals. The statement supports increased funding for enforcement of current regulations concerning the use of animals in biomedical research, encourages the development of nonanimal alternatives, and "encourages efforts to improve humane care of animals in both governmental and industrial research institutions. . . ."

Besch EL (1985) Definition of laboratory animal environmental conditions. In: Moberg GP (ed) *Animal stress.* Bethesda, MD: American Physiological Society, pp 297–315.
Reviews the current state of knowledge about optimal environmental conditions for common laboratory animals. Temperature and humidity, the microenvironment, ventilation, lighting, and noise are specifically addressed. The author advocates the use of voluntary guidelines to avoid the imposition of externally imposed restrictions. The concern is primarily for the validity of experimental results.

Crowley MA, Connors DD (1985) Critique of "The use of animals in nursing research." *Adv Nurs Sci 7*,4:23–31.
The authors present a critique of a 1982 article advocating wider use of animal experiments in nursing research (see Cunningham SG, Mitchell PH, 1982). Moral and scientific arguments are developed in opposition to the use of animals for experimentation, particularly in nursing research.

Culliton BJ (1985) HHS halts animal experiment. *Science 229*:447–448.
News and commentary about the July 18, 1985, announcement by HHS Secretary Margaret Heckler that she had suspended NIH funding for the University of Pennsylvania's head trauma laboratory. The article examines the role of activist groups in stealing videotapes from the laboratory, an action that initiated the NIH investigation of allegations of cruelty to baboons used in experiments there.

Food and Drug Administration. Agency Steering Committee on Animal Welfare Issues (1985) Final report to the acting commissioner Food and Drug Administration. *Clin Res Pract Drug Regul Aff 3*,2:231–260.
The Committee's recommendations include organizing workshops to address the following issues: acute toxicity studies, use of in vitro alternatives, and practices and procedures for the care and handling of animals. The committee also recommends the establishment of an agency-wide animal welfare committee with advisory responsibilities.

Fox JG (1985) Laboratory animal medicine—Changes and challenges. *Cornell Vet 75*,1:159–170.
Although this article is primarily concerned with the historical development of the profession of laboratory animal medicine rather than animal welfare issues, the author notes a connection between the two. Laboratory animal medicine is a specialty that is in the middle of the debate over animal rights—a challenge that the author believes will be met suc-

cessfully by assuaging public concern while implementing new regulations. The program from the first meeting (in 1950) of the American Association for Laboratory Animal Science (then called the Animal Care Panel) is reproduced, showing that one of the papers presented was entitled "Eradication of the Antivivisection Cult."

Frank M (1985a) "The lady doth protest too much, methinks." [editorial] *Physiol 28*,5:390.
The author, an officer of the American Physiological Society (APS), urges the scientific research community to launch a public relations campaign to convince the public of the medical benefits derived from animal research, and to ensure that institutional animal facilities are adequate and animals are not treated inhumanely. Otherwise, the author contends, raids by groups such as the Animal Liberation Front will continue to be considered justified by the public and by NIH. "The next time we cry foul, we cannot allow the animal rights groups to respond 'The lady doth protest too much, methinks.'"

Frank M (1985b) APS expresses concern; DHHS gives a reply. *Physiol 28*,5:408.
A letter written by Frank (American Physiological Society Executive VP) to Health and Human Services Secretary Margaret Heckler on behalf of APS, protesting Heckler's action in suspending federal funds for head trauma research at the University of Pennsylvania. The reply, written by James O. Mason, Acting Assistant Secretary for Health, denies Frank's claim that the DHHS "acced[ed] to the demands of demonstrators" and justifies the action on the grounds that "NIH amassed disturbing evidence that the practices for the care and use of these laboratory animals during recent years were materially out of compliance with the University's written assurances."

Gallup GG, Suarez SD (1985) Alternatives to the use of animals in psychological research. *Am Psychol 40*,10:1104–1111.
A rebuttal to the accusation that "scientists have been remiss in not developing alternatives to the use of animals in research." The authors assess various alternatives that have been proposed and conclude that "there are no viable alternatives to the use of live organisms in behavioral research." Includes a table presenting types and numbers of animals housed in graduate departments of psychology in the United States.

Gill TJ, III (1985) The rat in biomedical research. *Physiol 28*,1:9–17.
A survey of the contributions of the rat to biomedical and (the title notwithstanding) behavioral research. The author gives a detailed and rather technical overview of how rats have been used in studies of aging, autoimmune diseases, genetics, cancer, diabetes, cardiovascular phenomena, hypertension, infection, reproductive biology, transplantation, and behavioral science. Portrays rats as rather unpleasant creatures except in

their valuable role as research models. This is the first in a series on health benefits of animal research, later collected in a book: Gay W (ed) (1986) *Health benefits of animal research*, Washington, DC: Foundation for Biomedical Research. (See King FA, Yarbrough CJ, 1985.)

Gossell PP (1985) William Henry Welch and the antivivisection legislation in the District of Columbia, 1896–1900. *J Hist Med Allied Sci 40*,4:397–419.
A well-documented account of antivivisectionist ferment in the late 19th century in the United States, and particularly the role of Welch, Dean of Johns Hopkins Medical School, in defeating three antivivisectionist bills in the District of Columbia. The author explains events in terms of a struggle between traditional views of medicine and the new scientific medicine of the laboratory.

Greenwald GS (1985) ACDP survey on the use of animals in teaching physiology. *Physiol 28*,6:478–480.
Results of a survey undertaken by the Association of Chairmen of Departments of Physiology (ACDP) at the request of the Office of Technology Assessment (OTA) for its report to Congress. Answers to the 28 questions on animal use are presented in tabular form with brief comments. Major findings are that animal use has dropped over the past decade, but 84% of responding schools still have laboratories; of the latter, 89% use animals and 80% have dog labs.

Heidrich JE, Kent G (1985) Use of analgesics after surgery in animals. *J Am Vet Med Assoc 187*,5:513–514.
The authors urge the use of analgesics for animals after surgery and provide practical advice on selecting analgesics for dogs, cats, swine, ruminants, and horses. A table of recommended dosages of various nonnarcotic and narcotic analgesics is included for these species. The authors argue that one cannot assume an animal is not in pain "simply because in a similar situation human beings do not perceive pain."

Inglehart JK (1985) Health policy report: The use of animals in research. *N Engl J Med 313*,6:395–400.
A detailed discussion of legislative developments and recent activities of animal welfare proponents. The author presents both sides of the debate, praising "the fruits of medical research" as well as Congress's efforts to adopt "a tougher policy to ensure that animals are treated and cared for humanely. . . ." Particularly close attention is given to the 1985 revisions of the PHS policy (which would be revised again in 1986), and the role of NIH as it has evolved over the past several years.

Irving GW, III (1985) Regulations and guidelines for animal care: Problems and future concerns. In: Moberg GP (ed) *Animal stress*. Bethesda, MD: American Physiological Society, pp 281–296.

The author describes the basic provisions of the Animal Welfare Act (1966), the Good Laboratory Practice regulations (1978), NIH policies, and various bills introduced in the 98th Congress (1983–1984). Arguments are presented against any new legislation; better coordination of existing regulatory activities and educating the public are advocated instead. The premise is that the "improvement of science" should be the only consideration.

Jukes TH (1985) Opinions: Statement for panel discussion. *Physiol 28*,6:480–481.
The author views the animal rights movement as a severe threat to science and reminds readers that animal experiments led to the discovery of insulin as a treatment for diabetes. Dramatic photographs of diabetic patients before and after insulin treatment are included. Concludes that "the fight against the animal rights movement is not an argument about the size of cages but, rather, is part of a struggle by scientists to maintain the existence of science."

King FA, Yarbrough CJ (1985) Medical and behavioral benefits from primate research. *Physiol 28*,2:75–87.
This is the second in a series of articles on health benefits of animal research, justifying the use of animals in research, later collected in a book: Gay W (ed) (1986) *Health benefits of animal research*, Washington, DC: Foundation for Biomedical Research. (See Gill TJ III, 1985.) The authors offer an extensive survey of the ways in which nonhuman primates have been used in studies of atherosclerosis, aging, reproductive behavior, development of emotional expression (including depression), communication, infectious diseases, cancer, the brain, surgery and dentistry, and environmental health. Notes briefly a concern for the survival of threatened primate species.

King NB (1985) Council report: Benefits derived from research on food animals. *J Am Vet Med Assoc 186*,3:236–237.
The Council of the American Veterinary Medical Association (AVMA) reports its concern over the large number of bills introduced in 1984 "aimed at banning the use of animals for scientific research and restricting husbandry practices during the rearing and feeding process." The council believes that the use of animals for research to improve food production is justified, but alternatives should be used or developed where possible.

Kitchell RL, Johnson RD (1985) Assessment of pain in animals. In: Moberg GP (ed) *Animal stress*. Bethesda, MD: American Physiological Society, pp 113–140.
The authors define pain as a subjective perception and review an extensive literature on nociceptive mechanisms in humans and animals. The underlying premise is that "it can . . . be inferred that pain exists in humans as well as in animals unable to verbally communicate with humans.

Overemphasis on the uncertainty of knowledge about pain in animals that leads to a denial that pain perception exists in animals is logically as well as empirically unfounded."

Lederer SE (1985) Hideyo Noguchi's luetin experiment and the antivivisectionists. *Is 76*,281:31–48.
An account of the prolonged protests by antivivisectionists that followed the publication in 1911 of research done at the Rockefeller Institute in an effort to find a diagnostic test for syphilis. Noguchi used rabbits in the initial phases of the experiment, but the protests focused on his later use of human subjects as controls—46 healthy orphans and 190 adult patients hospitalized for diseases other than syphilis—without their consent. The antivivisectionists' concern over individual rights did not at that time result in legislation to regulate the use of human subjects in research, but it signaled the emergence of "a new emphasis on individual autonomy."

Miller NE (1985) The value of behavioral research on animals. *Am Psychol 40*,4:423–440.
A rebuttal to the accusation that behavioral research on animals is without value. The author cites a wide range of disorders for which therapeutic techniques have been developed from behavioral research using animals. Photos are included showing patients before and after treatment. "Who are the cruel and inhumane ones," the author asks, "the behavioral scientists whose research on animals led to the cures of the anorexic girl and the vomiting child [in the photos], or those leaders of the radical animal activists who are making an exciting career of trying to stop all such research and are misinforming people by repeatedly asserting that it is without any value?"

Mishkin B (1985) On parallel tracks: Protecting human subjects and animals. *Hastings Ctr Rep 15*,5:36–37.
Brief comparison of similarities between new Public Health Service (PHS) regulations affecting animal research and the regulations enacted 15 years ago to protect human research subjects. Judging from the experience with the latter, the author argues that animal researchers' fears about the new regulations are unfounded.

Overcast TD, Sales BD (1985) Regulation of animal experimentation. *J Am Med Assoc 254*,14:1944–1949.
The authors are lawyers who analyze the legal issues involved in the controversy between animal activists and the scientific community. Existing and proposed federal and state regulations are examined. The authors endorse weighing costs (to the animal) and benefits (to humans or other animals), but do not recommend writing these considerations into law. They conclude that further regulation "is neither necessary nor desirable." (Overcast has since publicly admitted that "substantial portions"

of this article were "taken word-for-word and sentence-for-sentence" from an earlier article by Chambers and Hines [see Chambers KT, Hines C (1983)]. Overcast assumed complete responsibility for this "horrendous but unintentional mistake" and offered four principles to help ensure the integrity of one's work [Overcast TD (1986) Regulation of animal experimentation: An apology. *J Am Med Assoc 255*,12:1568–1569].)

Prunieras M (1985) On the use of cell cultures in skin toxico- pharmacology. *Int J Dermatol 24*,2:98–100.
Reviews briefly the use of skin cultures in the field of toxicology in dermatology and cosmetology. The author urges further development of cell culture models to replace the large number of animals killed each year for drug testing and to replace human volunteers in drug tests.

Ritvo H (1985) Animal problems. Review of *Ethics and animals.* [book review] *Sci Tech Human Values 10*,3:87–91.
Review of Miller HB, Williams WH (eds) (1983) *Ethics and animals* (Clifton, NJ: Humana Press). The review provides a useful summary of the wide range of contributions in this volume, which grew out of a conference held at Virginia Polytechnic Institute and State University in 1979. The reviewer notes that the carefully reasoned, complex, and often ambivalent essays written by philosophers contrast sharply with the pragmatic pieces written by activists and suggests (with apparent regret) that philosophic deliberations will be ineffectual in determining the course of the public debate.

Rollin BE (1985) The moral status of research animals in psychology. *Am Psychol 40*,8:920–926.
A philosopher's thoughts on the subject. The author is critical of the status usually accorded research animals and argues that "no clear morally relevant differences exist to exclude animals from the moral arena." Some of the "special and difficult problems" raised by psychological research as opposed to biomedical research are noted. Because animals, like humans, have a *telos*—a nature, with physical and psychological needs and interests—the author concludes that "animals are morally more like children than like wheelbarrows."

Rosner F (1985) Is animal experimentation being threatened by animal rights groups? *J Am Med Assoc 254*,14:1942–1943.
Presents a broad overview of legislative initiatives since 1980, PHS policy regarding laboratory animals, the question of alternatives to animal use, and the response of the scientific community. The author does not give an unequivocal answer to the question posed in the title, but suggests steps scientists should take to defuse the issue.

Rowan AN, Goldberg AM (1985) Perspectives on alternatives to current animal testing techniques in preclinical toxicology. [review article] *Ann Rev Pharm Toxicol 25*:225–247.

The authors' premise is that animals are necessary in toxicology testing and will continue to be for a long time, but that there are many promising avenues for developing alternatives that could reduce the number of animals needed or could reduce their suffering. The latest developments in the following areas of testing are reviewed: LD–50, eye irritancy, dermal toxicity, liver toxicity, neurotoxicity, and fetal toxicity.

Samuels WM (1985a) People you should know . . . Doug Walgren. *Physiol 28*,1:5–7.
This was the first interview in what became a regular feature to introduce readers to people who have had or are likely to have an impact on physiology or on animal experimentation. Walgren (Democrat, PA) discusses the animal welfare legislation he has introduced in Congress and his reasons for believing such legislation to be necessary.

Samuels WM (1985b) People you should know . . . John McArdle. *Physiol 28*,2:69–72.
Interview with the director of laboratory animal welfare of the Humane Society of the United States (HSUS). Most of the HSUS positions described by McArdle are strongly opposed by the American Physiological Society (APS): for example, the HSUS's opposition to the release of pound animals for research, and support for more restrictive legislation. McArdle is not against *all* use of animals for research, however, and his call for increased appropriations for enforcement of the Animal Welfare Act is a position, Samuels notes, also taken by the APS.

Schumer W (1985) Editorial responsibility in the humane care and use of experimental animals. [editorial] *Circ Shock 15*,1:1–4.
The journal's editor puts prospective authors on notice that their research papers will henceforth be scrutinized for ethical and humane concerns. The Swedish system of classifying experiments (reproduced as a table) will be used to flag questionable papers. The editorial board's review of all papers published from 1974 to 1984 found 20 "questionable or incompletely described protocols."

Tannenbaum J, Rowan AN (1985) Rethinking the morality of animal research. *Hastings Ctr Rep 15*,5:32–43.
This is one of the most thoughtful and serious discussions of the philosophical issues of animal research to appear in the scientific literature to date. Critiques of the major ethical positions are offered, including analyses of the writings of Peter Singer, Tom Regan, Arthur Caplan, and Bernard Rollin. The authors find all of these ethical theories deficient and suggest principles that ought to be taken into account in developing an adequate theory. Recent empirical discoveries are brought into the discussion where relevant: for example, benzodiazepine receptors have been found in virtually all vertebrates, but not in invertebrates, indicating a physiological basis for imputing anxiety-related suffering to vertebrates,

but not to invertebrates. Although the authors do not present an ethical theory of their own, it is clear that they would support some, but not unrestricted, animal experimentation.

Wortman J (1985) PHS revises its lab animal welfare policy. *BioSci* *35*,7:408.
News feature reporting on the Public Health Service's (PHS) revised Policy on Humane Care and Use of Laboratory Animals by Awardee Institutions, which went into effect January 1, 1986. Outlines the new requirements that institutions receiving PHS grants must fulfill to be in compliance.

Wright EM Jr, Marcella KL, Woodson JF (1985) Animal pain: Evaluation and control. *Lab Anim* (NY) *14*,3:20–35.
Offers practical guidance to researchers in recognizing and alleviating pain in laboratory animals. Narcotic and nonnarcotic analgesic drugs and tranquilizers are discussed. The authors comment that it is curious that analgesic drugs are often withheld because it is not certain that the animals are experiencing pain, whereas "antibiotics are almost always given, even without documenting the presence of a bacterial infection. Pain and suffering may actually constitute the *only* situation in which one should go ahead and treat, even if in doubt."

Appendix A

The Animal Welfare Act
7 U.S.C., §2131–§2157

[Portions of this document that are italicized are the sections that were new or changed by the 1985 amendments.]
[PL 89–544, August 24, 1966, 80 Stat 350–353; as amended by PL 91–579, December 24, 1970, 84 Stat 1560–1565; PL 94–279, April 22, 1976, 90 Stat 417–423; and PL 99–198, title XVII, subtitle F, December 17, 1985, 99 Stat 1645–1650.]

Congressional Findings for 1985 Amendment:
Sec. 1751. For the purposes of this subtitle, the Congress finds that—

(1) the use of animals is instrumental in certain research and education for advancing knowledge of cures and treatment for diseases and injuries which afflict both humans and animals;

(2) methods of testing that do not use animals are being and continue to be developed which are faster, less expensive, and more accurate than traditional animal experiments for some purposes and further opportunities exist for the development of these methods of testing;

(3) measures which eliminate or minimize the unnecessary duplication of experiments on animals can result in more productive use of Federal funds; and

(4) measures which help meet the public concern for laboratory animal care and treatment are important in assuring that research will continue to progress.

Section 1. (a) This Act may be cited as the "Animal Welfare Act."

(b) The Congress finds that animals and activities which are regulated under this Act are either in interstate or foreign commerce or substantially affect such commerce or the free flow thereof, and that regulation of animals and activities as provided in this Act is necessary to prevent and eliminate burdens upon such commerce and to effectively regulate such commerce, in order—

(1) to insure that animals intended for use in research facilities or for exhibition purposes or for use as pets are provided humane care and treatment;

(2) to assure the humane treatment of animals during transportation in commerce; and

(3) to protect the owners of animals from the theft of their animals by preventing the sale or use of animals which have been stolen.

The Congress further finds that it is essential to regulate, as provided in this Act, the transportation, purchase, sale, housing, care, handling, and treatment of animals by carriers or by persons or organizations engaged in using them for research or experimental purposes or for exhibition purposes or holding them for sale as pets or for any such purpose or use.

Section 2. When used in this Act—

(a) The term "person" includes any individual, partnership, firm, joint stock company, corporation, association, trust, estate, or other legal entity;

(b) The term "Secretary" means the Secretary of Agriculture of the United States or his representative who shall be an employee of the United States Department of Agriculture;

(c) The term "commerce" means trade, traffic, transportation, or other commerce—

(1) between a place in a State and any place outside of such State, or between points within the same State but through any place outside thereof, or within any territory, possession, or the District of Columbia;

(2) which affects trade, traffic, transportation, or other commerce described in paragraph (1).

(d) The term "State" means a State of the United States, the District of Columbia, the Commonwealth of Puerto Rico, the Virgin Islands, Guam, American Samoa, or any other territory or possession of the United States;

(e) The term "research facility" means any school (except an elementary or secondary school), institution, organization, or person that uses or intends to use live animals in research, tests, or experiments, and that (1) purchases or transports live animals in commerce, or (2) receives funds under a grant, award, loan, or contract from a department, agency, or instrumentality of the United States for the purpose of carrying out research, tests, or experiments: *Provided,* That the Secretary may exempt, by regulation, any such school, institution, organization, or person that does not use or intend to use live dogs or cats, except those schools, institutions, organizations, or persons, which use substantial numbers (as determined by the Secretary) of live animals the principal function of which schools, institutions, organizations, or persons, is biomedical research or testing, when in the judgment of the Secretary, any such exemption does not vitiate the purpose of this Act;

(f) The term "dealer" means any person who, in commerce, for compensation or profit, delivers for transportation, or transports, except as a carrier, buys, or sells, or negotiates the purchase or sale of, (1) any dog or other animal whether alive or dead for research, teaching, exhibition, or use as a pet, or (2) any dog for hunting, security, or breeding purposes, except that this term does not include—

(i) a retail pet store except such store which sells any animals to a research facility, an exhibitor, or a dealer; or

(ii) any person who does not sell, or negotiate the purchase or sale of any wild animal, dog, or cat, and who derives no more than $500 gross income from the sale of other animals during any calendar year;

(g) The term "animal" means any live or dead dog, cat, monkey (nonhuman primate mammal), guinea pig, hamster, rabbit, or such other warm-blooded animal, as the Secretary may determine is being used, or is intended for use, for research, testing, experimentation, or exhibition purposes, or as a pet; but such term excludes horses not used for research purposes and other farm animals, such as, but not limited to livestock or poultry, used or intended for use as food or fiber, or livestock or poultry used or intended for use for improving animal nutrition, breeding, management, or production efficiency, or for improving the quality of food or fiber. With respect to a dog, the term means all dogs including those used for hunting, security, or breeding purposes;

(h) The term "exhibitor" means any person (public or private) exhibiting any animals, which were purchased in commerce or the intended distribution of which affects commerce, or will affect commerce, to the public for compensation, as determined by the Secretary, and such term includes carnivals, circuses, and zoos exhibiting such animals whether operated for profit or not; but such term excludes retail pet stores, organizations sponsoring and all persons participating in State and country fairs, livestock shows, rodeos, purebred dog and cat shows, and any other fairs or exhibitions intended to advance agricultural arts and sciences, as may be determined by the Secretary;

(i) The term "intermediate handler" means any person including a department, agency, or instrumentality of the United States or of any State or local government (other than a dealer, research facility, exhibitor, any person excluded from the definition of a dealer, research facility, or exhibitor, an operator of an auction sale, or a carrier) who is engaged in any business in which he receives custody of animals in connection with their transportation in commerce;

(j) The term "carrier" means the operator of any airline, railroad, motor carrier, shipping line, or other enterprise, which is engaged in the business of transporting any animals for hire;

(k) The term "Federal agency" means an Executive agency as such term is defined in section 105 of title 5, United States Code, and with respect to any research facility means the agency from which the research facility receives a Federal award for the conduct of research, experimentation, or testing, involving the use of animals;

(l) The term "Federal award for the conduct of research, experimentation, or testing, involving the use of animals" means any mechanism

(including a grant, award, loan, contract, or cooperative agreement) under which Federal funds are provided to support the conduct of such research.

(m) The term "quorum" means a majority of the Committee members;

(n) The term "Committee" means the Institutional Animal Committee established under section 13(b); and

(o) The term "Federal research facility" means each department, agency, or instrumentality of the United States which uses live animals for research or experimentation.

[For purposes of this Act, the term "animal" shall have the same meaning as defined in section 2(g) of the Animal Welfare Act (7 U.S.C. 2132(g)).]

Section 3. The Secretary shall issue licenses to dealers and exhibitors upon application therefore in such form and manner as he may prescribe and upon payment of such fee established pursuant to section 23 of this Act: *Provided,* That no such license shall be issued until the dealer or exhibitor shall have demonstrated that his facilities comply with the standards promulgated by the Secretary pursuant to section 13 of this Act: *Provided, however,* That any retail pet store or other person who derives less than a substantial portion of his income (as determined by the Secretary) from the breeding and raising of dogs or cats on his own premises and sells any such dog or cat to a dealer or research facility shall not be required to obtain a license as a dealer or exhibitor under this Act. The Secretary is further authorized to license, as dealers or exhibitors, persons who do not qualify as dealers or exhibitors within the meaning of this Act upon such persons' complying with the requirements specified above and agreeing, in writing, to comply with all the requirements of this Act and the regulations promulgated by the Secretary hereunder.

Section 4. No dealer or exhibitor shall sell or offer to sell or transport or offer for transportation, in commerce, to any research facility or for exhibition or for use as a pet any animal, or buy, sell, offer to buy or sell, transport or offer for transportation, in commerce, to or from another dealer or exhibitor under this Act any animal, unless and until such dealer or exhibitor shall have obtained a license from the Secretary and such license shall not have been suspended or revoked.

Section 5. No dealer or exhibitor shall sell or otherwise dispose of any dog or cat within a period of five business days after the acquisition of such animal or within such other period as may be specified by the Secretary: *Provided,* That operators of auction sales subject to section 12 of this Act shall not be required to comply with the provisions of this section.

Section 6. Every research facility, every intermediate handler, every carrier, and every exhibitor not licensed under section 3 of this Act shall register with the Secretary in accordance with such rules and regulations as he may prescribe.

Section 7. It shall be unlawful for any research facility to purchase any dog or cat from any person except an operator of an auction sale subject to section 12 of this Act or a person holding a valid license as a dealer or exhibitor issued by the Secretary pursuant to this Act unless such person is exempted from obtaining such license under section 3 of this Act.

Section 8. No department, agency, or instrumentality of the United States which uses animals for research or experimentation or exhibition shall purchase or otherwise acquire any dog or cat for such purposes from any person except an operator of an auction sale subject to section 12 of this Act or a person holding a valid license as a dealer or exhibitor issued by the Secretary pursuant to this Act unless such person is exempted from obtaining such license under section 3 of this Act.

Section 9. When construing or enforcing the provisions of this Act, the act, omission, or failure of any person acting for or employed by a research facility, a dealer, or an exhibitor or a person licensed as a dealer or an exhibitor pursuant to the second sentence of section 3, or an operator of an auction sale subject to section 12 of this Act, or an intermediate handler, or a carrier, within the scope of his employment or office, shall be deemed the act, omission, or failure of such research facility, dealer, exhibitor, licensee, operator of an auction sale, intermediate handler, or carrier, as well as of such person.

Section 10. Dealers and exhibitors shall make and retain for such reasonable period of time as the Secretary may prescribe, such records with respect to the purchase, sale, transportation, identification, and previous ownership of animals as the Secretary may prescribe. Research facilities shall make and retain such records only with respect to the purchase, sale, transportation, identification, and previous ownership of live dogs and cats. At the request of the Secretary, any regulatory agency of the Federal Government which requires records to be maintained by intermediate handlers and carriers with respect to the transportation, receiving, handling, and delivery of animals on forms prescribed by the agency, shall require there to be included in such forms, and intermediate handlers and carriers shall include in such forms, such information as the Secretary may require for the effective administration of this Act. Such information shall be retained for such reasonable period of time as the Secretary may prescribe. If regulatory agencies of the Federal Government do not prescribe requirements for any such forms, intermediate handlers and carriers shall make and retain for such reasonable period as the Secretary may prescribe such records with respect to the transportation, receiving, handling, and delivery of animals as the Secretary may prescribe. Such records shall be made available at all reasonable times for inspection and copying by the Secretary.

Section 11. All animals delivered for transportation, transported, purchased, or sold, in commerce, by a dealer or exhibitor shall be marked or identified at such time and in such humane manner as the Secretary may prescribe: *Provided*, That only live dogs and cats need be so marked or identified by a research facility.

Section 12. The Secretary is authorized to promulgate humane standards and recordkeeping requirements governing the purchase, handling, or sale of animals, in commerce, by dealers, research facilities, and exhibitors at auction sales and by the operators of such auction sales. The Secretary is also authorized to require the licensing of operators of auction sales where any dogs or cats are sold, in commerce, under such conditions as he may prescribe, and upon payment of such fee as prescribed by the Secretary under section 23 of this Act.

Section 13. (a) (1) The Secretary shall promulgate standards to govern the humane handling, care, treatment, and transportation of animals by dealers, research facilities, and exhibitors.

(2) The standards described in paragraph (1) shall include minimum requirements—

(A) for handling, housing, feeding, watering, sanitation, ventilation, shelter from extremes of weather and temperatures, adequate veterinary care, and separation by species where the Secretary finds necessary for humane handling, care, or treatment of animals; and

(B) for exercise of dogs, as determined by an attending veterinarian in accordance with general standards promulgated by the Secretary, and for a physical environment adequate to promote the psychological well-being of primates.

(3) In addition to the requirements under paragraph (2), the standards described in paragraph (1) shall, with respect to animals in research facilities, include requirements—

(A) for animal care, treatment, and practices in experimental procedures to ensure that animal pain and distress are minimized, including adequate veterinary care with the appropriate use of anesthetic, analgesic, tranquilizing drugs, or euthanasia;

(B) that the principal investigator considers alternatives to any procedure likely to produce pain to or distress in an experimental animal;

(C) in any practice which could cause pain to animals—

(i) that a doctor of veterinary medicine is consulted in the planning of such procedures;

(ii) for the use of tranquilizers, analgesics, and anesthetics;

(iii) for pre-surgical and post-surgical care by laboratory workers, in accordance with established veterinary medical and nursing procedures;

(iv) against the use of paralytics without anesthesia; and

(v) that the withholding of tranquilizers, anesthesia, analgesia, or euthanasia when scientifically necessary shall continue for only the necessary period of time;

(D) that no animal is used in more than one major operative experiment from which it is allowed to recover except in cases of—

(i) scientific necessity; or

(ii) other special circumstances as determined by the Secretary; and

(E) that exceptions to such standards may be made only when specified by research protocol and that any such exception shall be detailed and explained in a report outlined under paragraph (7) and filed with the Institutional Animal Committee.

(4) The Secretary shall also promulgate standards to govern the transportation in commerce, and the handling, care, and treatment in connection therewith, by intermediate handlers, air carriers, or other carriers, of animals consigned by any dealer, research facility, exhibitor, operator of an auction sale, or other person, or any department, agency, or instrumentality of the United States or of any State or local government, for transportation in commerce. The Secretary shall have authority to promulgate such rules and regulations as he determines necessary to assure humane treatment of animals in the course of their transportation in commerce including requirements such as those with respect to containers, feed, water, rest, ventilation, temperature, and handling.

(5) In promulgating and enforcing standards established pursuant to this section, the Secretary is authorized and directed to consult experts, including outside consultants where indicated.

(6) (A) Nothing in this Act—

(i) except as provided in paragraphs (7) of this subsection, shall be construed as authorizing the Secretary to promulgate rules, regulations, or orders with regard to the design, outlines, or guidelines of actual research or experimentation by a research facility as determined by such research facility;

(ii) except as provided subparagraphs (A) and (C) (ii) through (v) of paragraph (3) and paragraph (7) of this subsection, shall be construed as authorizing the Secretary to promulgate rules, regulations, or orders with regard to the performance of actual research or experimentation by a research facility as determined by such research facility; and

(iii) shall authorize the Secretary, during inspection, to interrupt the conduct of actual research or experimentation.

(B) No rule, regulation, order, or part of this Act shall be construed to require a research facility to disclose publicly or to the Institutional Animal Committee during its inspection, trade secrets or commercial or financial information which is privileged or confidential.

(7) (A) The Secretary shall require each research facility to show upon inspection, and to report at least annually, that the provisions of this Act are being followed and that professionally acceptable standards governing the care, treatment, and use of animals are being followed by the research facility during actual research or experimentation.

(B) In complying with subparagraph (A), such research facilities shall provide—

(i) information on procedures likely to produce pain or distress in any animal and assurances demonstrating that the principal investigator considered alternatives to those procedures;

(ii) assurances satisfactory to the Secretary that such facility is adhering to the standards described in this section; and

(iii) an explanation for any deviation from the standards promulgated under this section.

(8) Paragraph (1) shall not prohibit any State (or a political subdivision of such State) from promulgating standards in addition to those standards promulgated by the Secretary under paragraph (1).

(b) (1) The Secretary shall require that each research facility establish at least one Committee. Each Committee shall be appointed by the chief executive officer of each such research facility and shall be composed of not fewer than three members. Such members shall possess sufficient ability to assess animal care, treatment, and practices in experimental research as determined by the needs of the research facility and shall represent society's concerns regarding the welfare of animal subjects used at such facility. Of the members of the Committee—

(A) at least one member shall be a doctor of veterinary medicine;

(B) at least one member—

(i) shall not be affiliated in any way with such facility other than as a member of the Committee;

(ii) shall not be a member of the immediate family of a person who is affiliated with such facility; and

(iii) is intended to provide representation for general community interests in the proper care and treatment of animals; and

(C) in those cases where the Committee consists of more than three members, not more than three members shall be from the same administrative unit of such facility.

(2) A quorum shall be required for all formal actions of the Committee, including inspections under paragraph (3).

(3) The Committee shall inspect at least semiannually all animal study areas and animal facilities of such research facility and review as part of the inspection—

(A) practices involving pain to animals, and

(B) the condition of animals,

to ensure compliance with the provisions of this Act to minimize pain and distress to animals. Exceptions to the requirement of inspection of such study areas may be made by the Secretary if animals are studied in their natural environment and the study area is prohibitive to easy access.

(4) (A) The Committee shall file an inspection certification report of each inspection at the research facility. Such report shall—

(i) be signed by a majority of the Committee members involved in the inspection;

(ii) include reports of any violation of the standards promulgated, or assurances required, by the Secretary, including any deficient conditions of animal care or treatment, any deviations of research practices from originally approved proposals that adversely affect animal welfare, any notification to the facility regarding such conditions, and any corrections made thereafter;

(iii) include any minority views of the Committee; and

(iv) include any other information pertinent to the activities of the Committee.

(B) Such report shall remain on file for at least three years at the research facility and shall be available for inspection by the Animal and Plant Health Inspection Service and any funding Federal agency.

(C) In order to give the research facility an opportunity to correct any deficiencies or deviations discovered by reason of paragraph (3), the Committee shall notify the administrative representative of the research facility of any deficiencies or deviations from the provisions of this Act. If, after notification and an opportunity for correction, such deficiencies or deviations remain uncorrected, the Committee shall notify (in writing) the Animal and Plant Health Inspection service and the funding Federal agency of such deficiencies or deviations.

(5) The inspection results shall be available to Department of Agriculture inspectors for review during inspections. Department of Agriculture inspectors shall forward any Committee inspection records which include reports of uncorrected deficiencies or deviations to the Animal and Plant Health Inspection Service and any funding Federal agency of the project with respect to which such uncorrected deficiencies and deviations occurred.

(c) In the case of Federal research facilities, a Federal Committee shall be established and shall have the same composition and responsibilities provided in subsection (b), except that the Federal Committee shall report deficiencies or deviations to the head of the Federal agency conducting the research rather than to the Animal and Plant Health Inspection Service. The head of the Federal agency conducting the research shall be responsible for—

(1) all corrective action to be taken at the facility; and

(2) the granting of all exceptions to inspection protocol.

(d) Each research facility shall provide for the training of scientists, animal technicians, and other personnel involved with animal care and

treatment in such facility as required by the Secretary. Such training shall include instruction on—

(1) the humane practice of animal maintenance and experimentation;

(2) research or testing methods that minimize or eliminate the use of animals or limit animal pain or distress;

(3) utilization of the information service at the National Agricultural Library, established under subsection (e); and

(4) methods whereby deficiencies in animal care and treatment should be reported.

(e) The Secretary shall establish an information service at the National Agricultural Library. Such service shall, in cooperation with the National Library of Medicine, provide information—

(1) pertinent to employee training;

(2) which could prevent unintended duplication of animal experimentation as determined by the needs of the research facility; and

(3) on improved methods of animal experimentation, including methods which could—

(A) reduce or replace animal use; and

(B) minimize pain and distress to animals, such as anesthetic and analgesic procedures.

(f) In any case in which a Federal agency funding a research project determines that conditions of animal care, treatment, or practice in a particular project have not been in compliance with standards promulgated under this Act, despite notification by the Secretary or such Federal agency to the research facility and an opportunity for correction, such agency shall suspend or revoke Federal support for the project. Any research facility losing Federal support as a result of actions taken under the preceding sentence shall have the right of appeal as provided in sections 701 through 706 of title 5, United States Code.

(f)* No dogs or cats, or additional kinds or classes of animals designated by regulation of the Secretary, shall be delivered by any dealer, research facility, exhibitor, operator of an auction sale, or department, agency, or instrumentality of the United States or of any State or local government, to any intermediate handler or carrier for transportation in commerce, or received by any such handler or carrier for such transportation from any such person, department, agency, or instrumentality, unless the animal is accompanied by a certificate issued by a veterinarian licensed to practice veterinary medicine, certifying that he inspected the animal on a specified date, which shall not be more than ten days before such delivery, and, when so inspected, the animal appeared free of any infectious disease or physical abnormality which would endanger the animal or animals or other animals or endanger public health: *Provided, however,* That the Secretary may by regulation provide exceptions to this certifi-

*So in original. There are two subsections designated "(f)."

cation requirement, under such conditions as he may prescribe in the regulations, for animals shipped to research facilities for purposes of research, testing or experimentation requiring animals not eligible for such certification. Such certificates received by the intermediate handlers and the carriers shall be retained by them, as provided by regulations of the Secretary, in accordance with section 10 of this Act.

(g) No dogs or cats, or additional kinds or classes of animals designated by regulation of the Secretary, shall be delivered by any person to any intermediate handler or carrier for transportation in commerce except to registered research facilities if they are less than such age as the Secretary may by regulation prescribe. The Secretary shall designate additional kinds and classes of animals and may prescribe different ages for particular kinds or classes of dogs, cats, or designated animals, for the purposes of this section, when he determines that such action is necessary or adequate to assure their humane treatment in connection with their transportation in commerce.

(h) No intermediate handler or carrier involved in the transportation of any animal in commerce shall participate in any arrangement or engage in any practice under which the cost of such animal or the cost of the transportation of such animal is to be paid and collected upon delivery of the animal to the consignee, unless the consignor guarantees in writing the payment of transportation charges for any animal not claimed within a period of 48 hours after notice to the consignee of arrival of the animal, including, where necessary, both the return transportation charges and an amount sufficient to reimburse the carrier for all out-of-pocket expenses incurred for the care, feeding, and storage of such animals.

Section 14. Any department, agency, or instrumentality of the United States having laboratory animal facilities shall comply with the standards and other requirements promulgated by the Secretary for a research facility under sections 13 (a), (f), (g), and (h). Any department, agency, or instrumentality of the United States exhibiting animals shall comply with the standards promulgated by the Secretary under sections 13 (a), (f), (g), and (h).

Section 15. (a) The Secretary shall consult and cooperate with other Federal departments, agencies, or instrumentalities concerned with the welfare of animals used for research, experimentation or exhibition, or administration of statutes regulating the transportation in commerce or handling in connection therewith of any animals when establishing standards pursuant to section 13 and in carrying out the purposes of this Act. *The Secretary shall consult with the Secretary of Health and Human Services prior to issuance of regulations.* Before promulgating any standard governing the air transportation and handling in connection therewith, of animals, the Secretary shall consult with the Secretary of Transportation who shall have the authority to disapprove any such standard if he notifies the Secretary, within 30 days after such consultation, that changes in its provisions are necessary in the interest of flight safety. The Interstate Commerce Commission, the Civil Aeronautics Board, and the Federal Maritime Commission, to the extent of their respective lawful authorities, shall take such action as is appropriate to implement any standard established by the Secretary with respect to a person subject to regulation by it.

(b) The Secretary is authorized to cooperate with the officials of the various States or political subdivisions thereof in carrying out the purposes of this Act and of any State, local, or municipal legislation or ordinance on the same subject.

Section 16. (a) The Secretary shall make such investigations or inspections as he deems necessary to determine whether any dealer, exhibitor, intermediate handler, carrier, research facility, or operator of an auction sale subject to section 12 of this Act, has violated or is violating any provision of this Act or any regulation or standard issued thereunder, and for such purposes, the Secretary shall, at all reasonable times, have access to the places of business and the facilities, animals, and those records required to be kept pursuant to section 10 of any such dealer, exhibitor, intermediate handler, carrier, research facility, or operator of an auction sale. *The Secretary shall inspect each research facility at least once each year and, in the case of deficiencies or deviations from the standards promulgated under this Act, shall conduct such follow-up inspections as may be necessary until all deficiencies or deviations from such standards are corrected.* The Secretary shall promulgate such rules and regulations as he deems necessary to permit inspectors to confiscate or destroy in a humane manner any animals found to be suffering as a result of a failure to comply with any provision of this Act or any regulation or standard issued thereunder if (1) such animal is held by a dealer, (2) such animal is held by an exhibitor, (3) such animal is held by a research facility and is no longer required by such research facility to carry out the research, test, or experiment for which such animal has been utilized, (4) such animal is held by an operator of an auction sale, or (5) such animal is held by an intermediate handler or a carrier.

(b) Any person who forcibly assaults, resists, opposes, impedes, intimidates, or interferes with any person while engaged in or on account of the performance of his official duties under this Act shall be fined not more than $5,000, or imprisoned not more than three years, or both. Whoever, in the commission of such acts, uses a deadly or dangerous weapon shall be fined not more than $10,000, or imprisoned not more than ten years, or both. Whoever kills any person while engaged in or on account of the performance of his official duties under this Act shall be punished as provided under sections 1111 and 1114 of title 18, United States Code.

(c) For the efficient administration and enforcement of this Act, the provisions (including penalties) of sections 6, 8, 9, and 10 of the Act entitled "An Act to create a Federal Trade Commission, to define its powers and duties, and for other purposes," approved September 26, 1914 (38 Stat. 721–723, as amended; 15 U.S.C. 46, 48, 49, and 50) (except paragraph (c) through (h) of section 6 and the last paragraph of section 9), and the provisions of Title II of the "Organized Crime Control Act of 1970" (62 Stat. 856; 18 U.S.C. 6001 et seq.), are made applicable to the jurisdiction, powers, and duties of the Secretary in administering and enforcing the provisions of this Act and to any person, firm, or corporation with respect to whom such authority is exercised. The Secretary may prosecute any inquiry necessary to his duties under this Act in any part of the United States, including any territory, or possession thereof, the District of Columbia, or the Commonwealth of Puerto Rico. The powers conferred by said sections 9 and 10 of the Act of September 26, 1914, as amended, on the district courts of the United States may be exercised for the purposes of this Act by any district court of the United States. The United States district courts, the District Court of Guam, the District Court of the Virgin Islands, the highest court of American Samoa, and the United States courts of the other territories, are vested with jurisdiction specifically to enforce, and to prevent and restrain violations of this Act, and shall have jurisdiction in all other kinds of cases arising under this Act, except as provided in section 19 (c).

Section 17. The Secretary shall promulgate rules and regulations requiring dealers, exhibitors, research facilities, and operators of auction sales subject to section 12 of this Act to permit inspection of their animals and records at reasonable hours upon request by legally constituted law enforcement agencies in search of lost animals.

Section 18. Repealed by P.L. 91–579 (1970), which incorporated similar provisions in Section 13 (6) (A) (i).

Section 19. (a) If the Secretary has reason to believe that any person licensed as a dealer, exhibitor, or operator of an auction sale subject to section 12 of this Act, has violated or is violating any provision of this Act, or any of the rules or regulations or standards promulgated by the Secretary hereunder, he may suspend such person's license temporarily, but not to exceed 21 days, and after notice and opportunity for hearing, may suspend for such additional period as he may specify, or revoke such license, if such violation is determined to have occurred.

(b) Any dealer, exhibitor, research facility, intermediate handler, carrier, or operator of an auction sale subject to section 12 of this Act, that violates any provision of this Act, or any rule, regulation, or standard promulgated by the Secretary thereunder, may be assessed a civil penalty by the Secretary of not more than *$2,500* for each such violation, and the Secretary may also make an order that such person shall cease and desist from continuing such violation. Each violation and each day during which a violation continues shall be a separate offense. No penalty shall be assessed or cease and desist order issued unless such person is given notice and opportunity for a hearing with respect to the alleged violation, and the order of the Secretary assessing a penalty and making a cease and desist order shall be final and conclusive unless the affected person files an appeal from the Secretary's order with the appropriate United States Court of Appeals. The Secretary shall give due consideration to the appropriateness of the penalty with respect to the size of the business of the person involved, the gravity of the violation, the person's good faith, and the history of previous violations. Any such civil penalty may be compromised by the Secretary. Upon any failure to pay the penalty assessed by a final order under this section, the Secretary shall request the Attorney General to institute a civil action in a district court of the United States or other United States court for any district in which such person is found or resides or transacts business, to collect the penalty, and such court shall have jurisdiction to hear and decide any such action. Any person who knowingly fails to obey a cease and desist order made by the Secretary under this section shall be subject to a civil penalty of *$1,500* for each offense, and each day during which such failure continues shall be deemed a separate offense.

(c) Any dealer, exhibitor, research facility, intermediate handler, carrier, or operator of an auction sale subject to section 12 of this Act, aggrieved by a final order of the Secretary issued pursuant to this section may, within 60 days after entry of such an order, seek review of such order in the appropriate United States Court of Appeals in accordance with the provisions of section 2341, 2343 through 2350 of title 28, United States Code, and such court shall have exclusive jurisdiction to enjoin, set aside, suspend (in whole or in part), or to determine the validity of the Secretary's order.

(d) Any dealer, exhibitor, or operator of an auction sale subject to section 12 of this Act, who knowingly violates any provision of this Act shall, on conviction thereof, be subject to imprisonment for not more than 1 year, or a fine of not more than *$2,500*, or both. Prosecution of such violations shall, to the maximum extent practicable, be brought initially before United States magistrates as provided in section 636 of title 28, United States Code, and sections 3401 and 3402 of title 18, United States Code, and, with the consent of the Attorney General, may be conducted, at both trial and upon appeal to district court, by attorneys of the United States Department of Agriculture.

Section 20. Repealed by P.L. 94–279 (1976).*

Section 21. The Secretary is authorized to promulgate such rules, regulations, and orders as he may deem necessary in order to effectuate the purposes of this Act.

Section 22. If any provision of this Act or the application of any such provision to any person or circumstances shall be held invalid, the remainder of this Act and the application of any such provision to persons or circumstances other than those as to which it is held invalid shall not be affected thereby.

Section 23. The Secretary shall charge, assess, and cause to be collected reasonable fees for licenses issued. Such fees shall be adjusted on an equitable basis taking into consideration the type and nature of the operations to be licensed and shall be deposited and covered into the Treasury as miscellaneous receipts. There are hereby authorized to be appropriated such funds as Congress may from time to time provide: *Provided*, That there is authorized to be appropriated to the Secretary of Agriculture for enforcement by the Department of Agriculture of the provisions of section 26 of this Act an amount not to exceed $100,000 for the transition quarter ending September 30, 1976, and not to exceed $400,000 for each fiscal year thereafter.

Section 24. The regulations referred to in section 10 and section 13 shall be prescribed by the Secretary as soon as reasonable but not later than six months from the date of enactment of this Act. Additions and amendments thereto may be prescribed from time to time as may be necessary or advisable. Compliance by dealers with the provisions of this Act and such regulations shall commence ninety days after the promulgation of such regulations. Compliance by research facilities with the provisions of this Act and such regulations shall commence six months after the promulgation of such regulations, except that the Secretary may grant extensions of time to research facilities which do not comply with the standards prescribed by the Secretary pursuant to section 13 of this Act provided that the Secretary determines that there is evidence that the research facilities will meet such standards within a reasonable time. Notwithstanding the other provisions of this section, compliance by intermediate handlers, and carriers, and other persons with those provisions

*Section 20 of PL 89–544 (1966) provided civil penalties and provisions for appeal for research facilities that violated provisions of the Act. PL 94–279 (1976) repealed section 20 and incorporated its provisions in section 19, increasing the civil penalties for research facilities and adding criminal penalties for dealers, exhibitors, and operators of auction sales (but not for research facilities). PL 99–198 (1985) further increased the civil and criminal penalties incorporated in section 19. [Ed note]

of this Act, as amended by the Animal Welfare Act Amendments of 1976, and those regulations promulgated thereunder, which relate to actions of intermediate handlers and carriers, shall commence 90 days after promulgation of regulations under section 13 of this Act, as amended, with respect to intermediate handlers and carriers, and such regulations shall be promulgated no later than 9 months after the enactment of the Animal Welfare Act Amendments of 1976; and compliance by dealers, exhibitors, operators of auction sales, and research facilities with other provisions of this Act, as so amended, and the regulations thereunder, shall commence upon the expiration of 90 days after enactment of the Animal Welfare Act Amendments of 1976: *Provided, however,* That compliance by all persons with paragraphs (b), (c), and (d) of section 13 and with section 26 of this Act, as so amended, shall commence upon the expiration of said ninety-day period. In all other respects, said amendments shall become effective upon the date of enactment.

Section 25. Not later than March of each year following the enactment of the "Animal Welfare Act of 1970", the Secretary shall submit to the President of the Senate and the Speaker of the House of Representatives a comprehensive and detailed written report with respect to—

(1) the identification of all research facilities, exhibitors, and other persons and establishments licensed by the Secretary under section 3 and section 12 of this Act;

(2) the nature and place of all investigations and inspections conducted by the Secretary under section 16 of this Act, and all reports received by the Secretary under section 13 of this Act;

(3) recommendations for legislation to improve the administration of this Act or any provisions thereof; and

(4) recommendations and conclusions concerning the aircraft environment as it relates to the carriage of live animals in air transportation.

This report as well as any supporting documents, data, or findings shall not be released to any other persons, non-Federal agencies, or organizations unless and until it has been made public by an appropriate committee of the Senate or the House of Representatives.

Section 26. (a) It shall be unlawful for any person to knowingly sponsor or exhibit an animal in any animal fighting venture to which any animal was moved in interstate or foreign commerce.

(b) It shall be unlawful for any person to knowingly sell, buy, transport, or deliver to another person or receive from another person for purposes of transportation, in interstate or foreign commerce, any dog or other animal for purposes of having the dog or other animal participate in an animal fighting venture.

(c) It shall be unlawful for any person to knowingly use the mail service of the United States Postal Service or any interstate instrumentality for purposes of promoting or in any other manner furthering an animal fighting venture except as performed outside the limits of the States of the United States.

(d) Notwithstanding the provisions of subsections (a), (b), or (c) of this section, the activities prohibited by such subsections shall be unlawful with respect to fighting ventures involving live birds only if the fight is to take place in a State where it would be in violation of the laws thereof.

(e) Any person who violates subsection (a), (b), or (c) shall be fined not more than $5,000 or imprisoned for not more than 1 year, or both, for each such violation.

(f) The Secretary or any other person authorized by him shall make such investigations as the Secretary deems necessary to determine whether any person has violated or is violating any provision of this section, and the Secretary may obtain the assistance of the Federal Bureau of Investigation, the Department of the Treasury, or other law enforcement agencies of the United States, and State and local governmental agencies, in the conduct of such investigations, under cooperative agreements with such agencies. A warrant to search for and seize any animal which there is probable cause to believe was involved in any violation of this section may be issued by any judge of the United States or of a State court of record or by a United States magistrate within the district wherein the animal sought is located. Any United States marshal or any person authorized under this section to conduct investigations may apply for and execute any such warrant, and any animal seized under such a warrant shall be held by the United States marshal or other authorized person pending disposition thereof by the court in accordance with this paragraph (f). Necessary care including veterinary treatment shall be provided while the animals are so held in custody. Any animal involved in any violation of this section shall be liable to be proceeded against and forfeited to the United States at any time on complaint filed in any United States district court or other court of the United States for any jurisdiction in which the animal is found and upon a judgment of forfeiture shall be disposed of by sale for lawful purposes or by other humane means, as the court may direct. Costs incurred by the United States for care of animals seized and forfeited under this section shall be recoverable from the owner of the animals if he appears in such forfeiture proceeding or in a separate civil action brought in the jurisdiction in which the owner is found, resides, or transacts business.

(g) For purposes of this section—

(1) the term "animal fighting venture" means any event which involves a fight between at least two animals and is conducted for purposes of sport, wagering, or entertainment except that the term "animal fighting venture" shall not be deemed to include any activity the primary pur-

pose of which involves the use of one or more animals in hunting another animal or animals, such as waterfowl, bird, raccoon, or fox hunting;

(2) the term "interstate or foreign commerce" means—

(A) any movement between any place in a State to any place in another State or between places in the same State through another State; or

(B) any movement from a foreign country into any State;

(3) the term "interstate instrumentality" means telegraph, telephone, radio, or television operating in interstate or foreign commerce;

(4) the term "State" means any State of the United States, the District of Columbia, the Commonwealth of Puerto Rico, and any territory or possession of the United States;

(5) the term "animal" means any live bird, or any live dog or other mammal, except man; and

(6) the conduct by any person of any activity prohibited by this section shall not render such person subject to the other sections of this Act as a dealer, exhibitor, or otherwise.

(h) The provisions of this Act shall not supersede or otherwise invalidate any such State, local, or municipal legislation or ordinance relating to animal fighting ventures except in case of a direct and irreconcilable conflict between any requirements thereunder and this Act or any rule, regulation, or standard hereunder.

Section 27. (a) It shall be unlawful for any member of an Institutional Animal Committee to release any confidential information of the research facility including any information that concerns or relates to—

(1) the trade secrets, processes, operations, style of work, or apparatus; or

(2) the identity, confidential statistical data, amount or source of any income, profits, losses, or expenditures,

of the research facility.

(b) It shall be unlawful for any member of such Committee—

(1) to use or attempt to use to his advantages; or

(2) to reveal to any other person,

any information which is entitled to protection as confidential information under subsection (a).

(c) A violation of subsection (a) or (b) is punishable by—

(1) removal from such Committee; and

(2) (A) a fine of not more than $1,000 and imprisonment of not more than one year; or

(B) if such violation is willful, a fine of not more than $10,000 and imprisonment of not more than three years.

(d) Any person, including any research facility, injured in its business or property by reason of a violation of this section may recover all actual and consequential damages sustained by such person and the cost of the suit including a reasonable attorney's fee.

(e) Nothing in this section shall be construed to affect any other rights of a person injured in its business or property by reason of a violation of this section. Subsection (d) shall not be construed to limit the exercise of any such rights arising out of or relating to a violation of subsections (a) and (b).

Appendix B

The Health Research Extension Act of 1985, Public Law 99–158, November 20, 1985, "Animals in Research"

Sec. 495. (a) The Secretary, acting through the Director of NIH, shall establish guidelines for the following:

(1) The proper care of animals to be used in biomedical and behavioral research.

(2) The proper treatment of animals while being used in such research. Guidelines under this paragraph shall require—

(A) the appropriate use of tranquilizers, analgesics, anesthetics, paralytics, and euthanasia for animals in such research; and

(B) appropriate pre-surgical and post-surgical veterinary medical and nursing care for animals in such research.

Such guidelines shall not be construed to prescribe methods of research.

(3) The organization and operation of animal care committees in accordance with subsection (b).

(b)

(1) Guidelines of the Secretary under subsection (a)(3) shall require animal care committees at each entity which conducts biomedical and behavioral research with funds provided under this Act (including the National Institutes of Health and the national research institutes) to assure compliance with the guidelines established under subsection (a).

(2) Each animal care committee shall be appointed by the chief executive officer of the entity for which the committee is established, shall be composed of not fewer than three members, and shall include at least one individual who has no association with such entity and at least one doctor of veterinary medicine.

(3) Each animal care committee of a research entity shall—

(A) review the care and treatment of animals in all animal study areas and facilities of the research entity at least semiannually to evaluate compliance with applicable guidelines established under subsection (a) for appropriate animal care and treatment;

(B) keep appropriate records of reviews conducted under subparagraph (A); and

(C) for each review conducted under subparagraph (A), file with the Director of NIH at least annually (i) a certification that the review has been conducted, and (ii) reports of any violations of guidelines established under subsection (a) or assurances required under paragraph (1) which were observed in such review and which have continued after notice by the committee to the research entity involved of the violations.

Reports filed under subparagraph (C) shall include any minority views filed by members of the committee.

(c) The Director of NIH shall require each applicant for a grant, contract, or cooperative agreement involving research on animals which is administered by the National Institutes of Health or any national research institute to include in its application or contract proposal, submitted after the expiration of the twelve-month period beginning on the date of enactment this section—

(1) assurances satisfactory to the Director of NIH that—

(A) the applicant meets the requirements of the guidelines established under paragraphs (1) and (2) of subsection (a) and has an animal care committee which meets the requirements of subsection (b); and

(B) scientists, animal technicians, and other personnel involved with animal care, treatment, and use by the applicant have available to them instruction or training in the humane practice of animal maintenance and experimentation, and the concept, availability, and use of research or testing methods that limit the use of animals or limit animal distress; and

(2) a statement of the reasons for the use of animals in the research to be conducted with funds provided under such grant or contract.

Notwithstanding subsection (a)(2) of section 553 of title 5, United States Code, regulations under this subsection shall be promulgated in accordance with the notice and comment requirements of such section.

(d) If the Director of NIH determines that—

(1) the conditions of animal care, treatment, or use in an entity which is receiving a grant, contract, or cooperative agreement involving research on animals under this title do not meet applicable guidelines established under subsection (a);

(2) the entity has been notified by the Director of NIH of such determination and has been given a reasonable opportunity to take corrective action; and

(3) no action has been taken by the entity to correct such conditions;

the Director of NIH shall suspend or revoke such grant or contract under such conditions as the Director determines appropriate.

(e) No guideline or regulation promulgated under subsection (a) or (c) may require a research entity to disclose publicly trade secrets or commercial or financial information which is privileged or confidential.

APPENDIX C

Interdisciplinary Principles and Guidelines for the Use of Animals in Research, Testing, and Education

Introduction

The *Interdisciplinary Principles and Guidelines for the Use of Animals in Research, Testing, and Education* were developed by The New York Academy of Sciences' Ad Hoc Committee on Animal Research in conjunction with an Advisory Panel. They represent the culmination of an extensive effort by scientists, educators, administrators, and members of the general public.

Our objective was to establish an ethical and educational approach to the care and treatment of animals, acceptable across many scientific disciplines, for students, scientists, and others involved in any aspect of experimentation, testing, or educational projects using animals. It is our hope that these principles and guidelines will provide a standard for animal use and will be valuable to organizations as they develop or expand their existing guidelines. It is also our hope that these guidelines will be used as the basis of educational programs.

Acknowledgments

The Ad Hoc Committee on Animal Research is grateful to the Board of Governors of The New York Academy of Sciences for their support and encouragement. We also thank the Academy staff for their special assistance in helping to bring this project to completion.

We thank the members of the Advisory Panel and all others who contributed to these principles and guidelines but in particular we are indebted to Professor Neal Miller for his invaluable guidance, insights, and knowledge of animal experimentation across the broad spectrum of sci-

---

Prepared by the New York Academy of Sciences' Ad Hoc Committee on Animal Research. Appendix C is reprinted with permission of the New York Academy of Sciences.

ence; to Professor Edward Gershey for his succinct analysis which appears here as Appendix II; to the American Veterinary Medical Association for their assistance in preparing Appendix III; to Professor Roger Greif for his tireless efforts in editing the many drafts of this document; and to Professor Sir William Paton for his illuminating comments which extended the breadth and depth of our project.

Gratitude is expressed to all the organizations that provided financial support for this project. Finally, we thank Dr. Norbert Roberts and Dr. Robert Scala who continued to encourage us from the time we first discussed this project with them several years ago.

ADVISORY PANEL

CONSULTANTS

Congressman Bill Green, (R-NY), Washington, D.C.
Adele Douglas, Legislative Aide to Congressman Green, Washington, D.C.

Arthur L. Caplan, Ph.D., The Hastings Center, Briarcliff Manor, New York

James Mahoney, D.V.M., New York University Medical Center, Tuxedo, New York

REPRESENTATIVES

Robert Drew, Ph.D., American Petroleum Company, Washington, D.C. (Society of Toxicology)

Bruce H. Ewald, M.A., D.V.M., Ciba-Geigy Corp., Summit, New Jersey, (Ciba-Geigy Corp.)

Ronald Dubner, D.D.S., Ph.D., National Institute of Dental Research, NIH, Bethesda, Maryland (American Pain Society, International Association for the Study of Pain)

Henry Edinger, Ph.D., J.D., University of Medicine and Dentistry of New Jersey, Newark, New Jersey (Society for Neuroscience)

Edward Gershey, Ph.D., The Rockefeller University, New York, New York (The New York Academy of Medicine)

Roger Greif, M.D., Cornell University Medical College, New York, New York (The Endocrine Society)

Mimi Halpern, Ph.D., SUNY Health Science Center, Brooklyn, New York (American Association of Anatomists)

Joseph Harun, M.D., Carter-Wallace, Inc., Cranbury, New Jersey (Carter-Wallace, Inc.)

Joseph Haught, D.V.M., Schering-Plough Corp., Bloomfield, New Jersey (Schering-Plough Corp.)

David Hysell, D.V.M., Procter & Gamble Company, Cincinnati, Ohio (Procter & Gamble Company)

Francis A. Kallfelz, D.V.M., New York State College of Veterinary Medicine, Ithaca, New York (American Veterinary Medical Association)

John Laszlo, M.D., American Cancer Society, New York, New York (American Association for Cancer Research)

Leonard Lerner, Ph.D., Jefferson Medical College, Philadelphia, Pennsylvania (Society for the Study of Reproduction)

Neal E. Miller, Ph.D., Yale University, New Haven, Connecticut (American Psychological Association)

Robert A. Scala, Ph.D., Exxon Biomedical Sciences, Inc., East Millstone, New Jersey (Exxon Corp.)

Leonard T. Search, Schering-Plough Corp., Lafayette, New Jersey (Schering-Plough Corp.)

Lawrence R. Soma, V.M.D., University of Pennsylvania, School of Veterinary Medicine, Philadelphia, Pennsylvania (American College of Veterinary Anesthesia)

Stephen Springer, Ph.D., Vicks Research Center, Shelton, Connecticut (Richardson-Vicks, Inc.)

Frank Vymetal, D.V.M., Schering-Plough Corp., Lafayette, New Jersey (Schering-Plough Corp.)

Judith Weis, Ph.D., Rutgers University, Newark, New Jersey (American Society of Zoologists)

John Worsing, D.V.M., American Cyanamid Company, Pearl River, New York (American Cyanamid Company)

AD HOC COMMITTEE ON ANIMAL RESEARCH

Chair: Jeri A. Sechzer, Ph.D., Cornell University Medical College, New York, New York

Doreen Berman, Ph.D., Queens College, City University of New York, Flushing, New York

Craig Burrell, M.D., Sandoz Corp., East Hanover, New Jersey

Barbara P. Carter, New York, New York

Bruce H. Ewald, M.A., D.V.M., Ciba-Geigy Corp., Summit, New Jersey

Nancy Geller, Ph.D., Memorial Sloan-Kettering Cancer Center, New York, New York

Anne Griffin, Ph.D., The Cooper Union for the Advancement of Science and Art, New York, New York

Phyllis B. Grodsky, Ph.D., Memorial Sloan-Kettering Cancer Center, New York, New York

Don Grubman, J.D., New York, New York

Leon Lewis, D.V.M., Hoffmann-La Roche, Inc., Nutley, New Jersey

Jacqueline Messite, M.D., New York Academy of Medicine, New York, New York

Brian Morgan, Ph.D., Columbia University College of Physicians and Surgeons, New York, New York

Elizabeth Muchmore, M.D., New York University Medical Center, Tuxedo, New York

David D. Myers, D.V.M., Ph.D., Memorial Sloan-Kettering Cancer Center and Cornell University Medical College, New York, New York

Robert A. Scala, Ph.D., Exxon Biomedical Sciences, Inc., East Millstone, New Jersey

Philip H. Sechzer, M.D., SUNY Health Science Center, Brooklyn, New York

Philip Siekevitz, Ph.D., The Rockefeller University, New York, New York

Lawrence R. Soma, V.M.D., University of Pennsylvania, School of Veterinary Medicine, Philadelphia, Pennsylvania

Dennis M. Stark, D.V.M., Ph.D., The Rockefeller University, New York, New York

Robert L. Thompson, Ph.D., Hunter College, City University of New York, New York, New York

---

FINANCIAL ASSISTANCE

The Ad Hoc Committee on Animal Research gratefully acknowledges the financial support of the following organizations:

- Bristol-Myers Company
- Carter-Wallace Incorporated
- Colgate-Palmolive Company
- Dow Chemical Company
- Exxon Education Foundation
- Merck, Sharpe, & Dohme Laboratories
- Procter & Gamble Company
- Richardson-Vicks Incorporated
- Schering-Plough Foundation
- Texaco Philanthropic Foundation

Interdisciplinary Principles and Guidelines for the Use of Animals in Research, Testing, and Education

The scientific community seeks to understand, improve, and sustain life and alleviate suffering. The pursuit of these goals may necessitate the use of animals when new information is sought and older information is to be reassessed, validated, or demonstrated for educational purposes, and alternative methods are unavailable. Concern for the welfare of animals used in research, testing, and education has led scientists to adopt more comprehensive standards so that gains in knowledge can be achieved using the smallest number of animals and the most humane methods.

The guiding principles which follow should be carefully considered in the development of the experimental design and the establishment of all procedures to assure the most humane treatment of animals possible without compromising scientific validity. These guidelines outline some of the major concerns regarding humane treatment of animals, but are by no means an exhaustive discourse on all possible circumstances. They are intended to guide the conduct of investigators, teachers, students, and research personnel and can serve as a reference for Institutional Animal Care and Use Committees, editors reviewing scientific manuscripts for publication, and others interested in current standards. They may be cited in conjunction with publication of all experiments involving animals.

These principles and guidelines were developed by an Ad Hoc Committee of The New York Academy of Sciences with input from individuals representing a wide spectrum of disciplines, scientific societies, and organizations. They are intended to be applicable to all research, testing, and educational projects regardless of the source of funding. They are intended to be in agreement with current Public Health Service Policy and Federal law.

The Use of Animals in Research and Testing

The Institutional Animal Care and Use Committee (IACUC)

Each organization which uses animals for research, testing, or education should establish an Institutional Animal Care and Use Committee (IACUC) to review and approve all proposals related to the use of animals. The IACUC has the responsibility of monitoring compliance by investigators with institutional, governmental, and these guidelines with respect to the humane care and treatment of animals. The committee should consist of at least five (5) persons and should include one veterinarian, one practicing scientist, one person unaffiliated with the organization, and a non-scientist.[1] It should also include or seek the advice of specialists in the techniques used at the organization (e.g., anesthesiologists, statisticians, etc.). The committee should develop a procedure within the scope of regulatory requirements for reviewing proposals and dealing with concerns of and appeals by investigators.

A guide to the selection of appropriate animal procedures is presented in Appendix I and a sample of a protocol for reviewing proposals is given in Appendix II. Proposals may be approved by the IACUC chair or a committee member so designated by the chair, provided that all committee members have been informed of the proposals under review and provided that any committee member may request full committee review. *Full committee* review consists of review at a convened meeting where a quorum is present. Approval by the full committee consists of an affirmative vote by a majority of those present.

Individual repetitive testing procedures, using a standardized protocol previously reviewed by the committee, need not be reviewed again unless a procedure is modified and has a significantly different effect on the animal. Ongoing projects should be reviewed at appropriate intervals, but not less than once every three years to assure the humane treatment of animals. The IACUC should periodically, but at least semiannually, inspect not only housing facilities, but also areas where research is being conducted, especially when experiments are long term, of Types F and G in Appendix I, or where survival surgery is performed.

The IACUC should ensure that there is an institutional mechanism to provide the required initial and continuing educational programs and activities for all individuals involved in the use of animals for research, testing, and education.[2]

Animal Model Selection

In planning a procedure, the investigator/instructor should first consider whether non-animal methods (e.g., audio-visual, tissue culture) could *replace* the use of animals. When animals are needed, thought should be given first to the use of the least complex species required to answer the given question; second, to the use of protocols or technologies that *reduce* to a minimum the number of animals needed; and third, where pain or distress is likely, to *refine* existing procedures and techniques so that the level of pain or distress to the animal, regardless of species, is reduced to the practicable minimum necessary for the purpose of the research.

Animals bred specifically for laboratory study should be used wherever practicable. By using known strains, data can be more readily compared and responses will be more predictable. However, there are situations where the use of random source animals may be appropriate.

In planning animal procedures, consideration should be given to equal use of females and males. At present, much research is conducted on male animals; their female littermates, not used for breeding, frequently are disposed of. Results of studies using only males have often been assumed to apply to females as well, sometimes without justification. The appropriate use of both male and female animals may increase the validity of data and decrease animal wastage.

Regulatory and Legal Considerations[3]

Only animals that are lawfully acquired should be used in the laboratory or classroom and their care and use should be in compliance with federal, state, and local laws and regulations regardless of the source of funding.

Replication and Duplication of Experiments

Replication *per se* is a critical part of the scientific method. Results of an experiment can be considered valid only if they can be reproduced in other experiments and/or by other scientists. Without such procedures, the possibility of experimental error cannot be eliminated. Moreover, replication is a method of "checking" results which is conservative of animal life since it helps prevent the use of erroneous data as a basis for additional studies using animals. Replication experiments should be planned as carefully as new experiments in order to minimize animal use.

To reduce unnecessary/unintentional duplication in research, each investigator is encouraged to conduct a comprehensive literature search as part of planning every new research proposal. Information regarding the latest applicable non-animal technology and/or methodology should be sought and should be used, wherever possible, to replace, in whole or in part, the use of live animals.

Responsibilities of Research Staff

Procedures on animals should be carried out only by or under the supervision of investigators/instructors with appropriate training and experience in biological and/or behavioral sciences. It is the responsibility of investigators/instructors to ensure that those under their supervision have received explicit instruction in experimental methods and in the care, maintenance, and handling of the species being studied. It is also the responsibility of the investigator/instructor to supply to the IACUC complete details of experiments which involve the use of live animals. Responsibilities and activities of all individuals dealing with animals shall be consistent with their respective skills, training, and experience.

Avoidance or Minimization of Pain or Distress

When designing an experiment, the investigator/instructor should consider the degree of pain or distress, if any, it will cause the animals.[4] The greater the pain and distress a procedure is likely to cause an animal, the more important it is that the experimental design is sound and that due consideration has been given to reducing any unnecessary pain or distress. Appendix I to these guidelines can help investigators classify the potential of procedures for causing pain or distress and determine the measures necessary to keep pain or distress at a minimum. Appendix II will help investigators by raising specific questions regarding the design of their experiments and the appropriate use of these guidelines.

The design and conduct of toxicity or product safety evaluation studies deserves special mention. The fact that these experiments may be mandated by law or regulation does not decrease the need for sensitivity to unnecessary pain or distress.

Anesthesia, analgesia, or sedation should be used to reduce pain or distress unless there is scientific justification that their use is incompatible with the aim of the procedure.

If it is necessary to carry out procedures on unanesthetized animals which are expected to cause more than minimal pain or distress, the procedure should be reviewed by the full IACUC. The maximum degree, duration, and frequency of pain or distress to which the animals will be exposed should be specified by the investigator as should criteria for determining the signs of pain or distress in the species used. Although pain and distress may be perceived differently in different animal species, severe pain or distress is to be construed as that which human beings would consider intolerable without palliative measures.[5]The following subsections highlight specific issues concerning pain and distress which arise in the conduct of biological and/or behavioral research. *When the purpose of a study conflicts with these guidelines, the detailed protocol should be submitted and approved by the full IACUC.*

Surgical Procedures

To prevent pain, surgical procedures may only be conducted on animals which have received sufficient general or regional anesthesia with appropriate agents. Food should be withheld from animals prior to anesthesia to assure the extent of stomach emptying necessary to prevent aspiration of gastric contents. Since there are species differences in gastric emptying time, duration of food deprivation should be determined according to the species used.

Surgery may only be performed under aseptic techniques appropriate to the species except in experiments of Type D in Appendix I in which the animal will not be permitted to recover from anesthesia. If the intent is to allow the animal to recover from anesthesia, then appropriate aseptic precautions must be taken to prevent infection.[6]

Post-surgical analgesia must be provided appropriate for the type of surgical intervention. Appropriate post-operative care must also be provided to assure adequate wound healing, to prevent undesirable effects of surgery, and to minimize post-operative discomfort.

When a goal of an experiment is assessment of pain or distress occurring as a result of surgical intervention, and therefore pain-alleviating agents cannot be employed, the investigator must specify the appropriate endpoint at which the experiment will be terminated or the animal humanely euthanized.

Studies of Drugs or Other Chemical Agents

Unlike many studies in which the impact of the procedure on the experimental animal is often known in advance (e.g., Appendix I, Types F and G), studies of drugs or other chemical agents to determine pharmacological and/or toxicological outcomes are often unpredictable. For such studies, standardized protocols, which have been reviewed by the full IACUC, should be in place and general criteria for euthanasia or termination of the experiment on humane grounds should be understood by all participants in the study.

Altering the State of the Animal

In cases where an animal's state of awareness or sensory perception is altered, particularly after the use of analgesics, anesthetics, or other pharmacologically active substances, as part of certain toxicity studies, or after surgical procedures, serious attempts should be made to protect the animal from accidental or self-inflicted injury.

When studies involve psychological and physiological dependence on drugs, special procedures should be planned to avoid or to minimize withdrawal syndromes unless withdrawal is the specific phenomenon under study or unless such procedures will interfere with the goals of the study.

Use of Behavioral Techniques

Behavioral techniques or behavioral modifications may be employed:

a. Specifically for the study of behavior, including abnormal behavior,
b. For studies in various disciplines to elucidate particular mechanisms or functions,

c. To socialize an animal with regard to its human handlers or,
d. To adapt an animal to a new environment or rehabituate an animal to a previous environment.

These procedures may range from those that involve no pain or distress to those which expose the animal to more than moderate pain or distress, examples of which are given in Appendix I to these guidelines.

In general, appetitive motivation is preferable to aversive stimulation when a choice is feasible. When an aversive procedure is necessary, investigators/instructors are urged to design the procedure for the least aversive combination of intensity, duration, frequency, and type of stimulus compatible with the goals of the study.

Whenever possible, procedures should allow the animal to control, alleviate, avoid, or escape from aversive stimuli, or to self-administer analgesia, particularly in procedures involving pain or distress. Any exceptions to allowing the animal such control should be approved by the full IACUC.

Multiple-Use Considerations

An animal which has been used in major survival surgery or in a procedure involving more than minimal pain or distress should not be used in additional procedures unless the additional procedure is of scientific necessity and is an integral part of the same experiment or will not involve more than minimal pain or distress, or unless the animal is anesthetized throughout the procedure and not allowed to recover, or unless the species required is in such short supply that multiple use is necessary.

Necessity of Chemical Restraint or Paralysis

Neuromuscular-blocking agents which cause motor paralysis without loss of sensation are not anesthetics and, when required, should be used only with an anesthetic, not in place of it. Since motor paralysis increases the difficulty of evaluating the depth of anesthesia, experienced personnel and proper equipment or procedures *must* be available throughout the procedure to monitor physiological signs and assure that the animal remains anesthetized. Neuromuscular-blocking agents without anesthesia should only be used in exceptional cases where there is no alternative and only after approval by the full IACUC.

Evaluation of Physical Restraint

Prolonged physical restraint as part of an experiment should be considered only when alternative procedures are inadequate. Prior to using a particular type of restraint for the first time, investigators should consult with a person knowledgeable about the effects of different types of restraint on the particular species in the study. It is essential to ensure that restraint is of the minimal degree and duration necessary, not only to limit the distress to the animal, but because significant amounts of distress may affect the results of an experiment, reducing validity and wasting an animal resource.

Convenience of the investigator/instructor is not a justification for using prolonged restraint. Such restraint should be carried out only to achieve the scientific goal of the experiment with the approval of the full IACUC. Whenever restraint is used, the animal should be attended frequently to avoid the possible undesirable effects of these procedures.

Alteration of Life Support Systems

The life support system of the animal should be of particular concern in the design and conduct of certain experimental procedures. Variables which may modify or alter life support may interfere with the animal's well being and affect experimental results. The following subsections address some of these variables.

Food Deprivation

When it is necessary to withhold food from an animal, the following conditions should be considered:

a. Consideration should be given to the possibility of shortening or eliminating food deprivation as motivation by using a highly preferred food as an incentive.
b. When nutrition is not the objective of the study, special attention should be given to ensuring a suitably balanced diet for the animal.
c. In short term procedures which involve withholding food, except prior to anesthesia, the investigator should provide information to the IACUC on the consequences of food deprivation of the extent proposed. If the period of deprivation will place the animal's health and well-being in jeopardy, the procedure must be approved by the institution's full IACUC.
d. In long term studies which require body weight reduction of both immature or mature animals, the reduction should be known to have no overt detrimental effects on the animal. For commonly used mammalian species it is recommended that the decrease in body weight be no greater than 20% of the pre-experimental weight of the animal to be deprived or 20% of the mean weight of appropriate control animals. For deprivation beyond this guideline, or when detrimental effects of the deprivation are the main purpose of the study, approval of the full IACUC is required.
e. In long term studies involving deprivation or excess of one or a combination of essential nutrients in both immature or mature animals, consideration must be given to the metabolic and physical conse-

quences to the animal and steps must be taken to promptly detect and alleviate pain or distress resulting from such experimental procedures. When such studies may result in conditions which, of themselves, are painful or distressful (e.g., ulcerations, fractures, and infections), approval by the full IACUC is required.

Water Deprivation

Water deprivation should be used primarily when thirst or drinking is the specific object of the study or if required for the safety of the animal prior to anesthetic procedures. When water deprivation is contemplated for other procedures, special consideration must be given since water deprivation usually affects an animal's physiological and metabolic condition more than does food deprivation. The period of withholding fluid should not ordinarily exceed limits similar to those established by the U.S. Department of Agriculture for the transport of the species concerned. If these are exceeded, the approval of the full IACUC is required.[7]

Sleep Deprivation

If an animal is to be deprived of sleep, the proposal must give details of the methods to be used and of the distress likely to be caused by them. An assessment should also be made of the possible physiological consequences of sleep deprivation on animals in the study.

Significant Environmental Alterations

Studies involving significant modification of gravitational fields, hibernation patterns, temperature, light, noise, humidity, atmospheric pressure, levels of oxygen and other gases, as well as other environmental variables, should be carried out only after approval by the institution's full IACUC. Consideration should be given to species-specific sensitivity to environmental conditions.

Euthanasia

Euthanasia must be considered a major responsibility. It is that method of killing an animal which assures minimal physical and psychological suffering. Euthanasia should be carried out in compliance with those methods currently recommended by the American Veterinary Medical Association (AVMA) (see Appendix III).

a. Whenever it is necessary to euthanize an animal, the procedure should be conducted in accordance with accepted guidelines for the humane killing of experimental animals and with regard to the collection or analysis of specimens. *When, for scientific reasons, the procedure cannot be conducted in accordance with AVMA guidelines, it should be regarded as part of the experimental procedure with specific approval by the full IACUC.*

b. If, for any reason, unrelated to the aims of the study, an animal suffers unrelievable, severe, acute, or persistent pain, distress, or disablement, it should be euthanized promptly by the investigator or someone authorized by the investigator.
c. For studies in which death of animals is an inevitable part of the protocol, every effort should be made to determine an acceptable endpoint when euthanasia can be performed to prevent unnecessary suffering (see Appendix I, Type G).
d. An animal must not be discarded until its death is verified by criteria appropriate to the experiment.
e. After euthanasia, animal organs and tissues should be shared among investigators/instructors wherever feasible to reduce further animal use.

Disposition of Animals

Animals used in procedures involving only minimal pain or distress may be offered to other investigators/instructors for additional study provided these guidelines are followed. A protocol for such use should be reviewed and approved by the IACUC.

If, for any reason, laboratory animals are to be placed in a non-laboratory setting, consideration should be given to their adaptation to a new environment or rehabituation to their natural environment. However, adaptation and especially rehabituation should not be undertaken without prior consideration of the myriad potential problems such as social readjustment, competition with other resident species, impact on the environment of overpopulation, destruction of vegetation, susceptibility to diseases endemic in the area and, conversely, the susceptibility to disease of the introduced animal. The decision to proceed with adaptation or rehabituation should be made with the advice of and participation of appropriate experts such as veterinarians, animal behaviorists, or ecologists.

The Use of Animals in Education

Colleges and Universities, Graduate and Professional Schools

There are situations in which currently there is no substitute for the use of living animals for educational purposes. When such use is planned, the need to use live animals should be justified in a proposal. All proposals involving live animals in educational projects should be reviewed and approved by the IACUC. The procedure should always involve the minimum number of animals and should be designed and conducted in accordance with the previously stated guidelines.

The IACUC, in conjunction with the department concerned, should also advise their administration on the need to establish a policy to deal with concerns of students who wish to absent themselves from participating in classroom experiments involving live animals.

1. Videotapes and films should be used wherever feasible if they can replace the use of live animals without loss of educational value. This particularly applies to demonstrations of known phenomena and those procedures likely to cause the animal pain, distress, or disablement.
2. Undergraduate students should not personally perform procedures likely to cause an animal more than momentary pain or distress without consideration by the department and the IACUC of the relevance of the procedure for the aims of the curriculum and the need for supervision.
3. Curricula in academic disciplines that require animal experiments should include material on:

   a. The history of animal use, current ethical issues in animal research, and alternatives to the use of animals,
   b. The care and use of laboratory animals,
   c. The effects of the laboratory environment and stressors on the animal and on subsequent experimental results.

4. Any student preparing to use animals in classroom experiments should be instructed in the handling and care of animals. Students should be supervised by someone with extensive experience until it is well demonstrated that they have the ability to carry out these experiments.
5. Whenever animals are used for instructional purposes, every attempt should be made to limit procedures to Types A through C of Appendix I.

   a. Demonstration should be encouraged in order to conserve animals. If feasible, several students should work with one animal rather than have one animal for each student.
   b. If it is considered necessary to conduct a procedure involving Types D through G of Appendix I, a veterinarian should be consulted in the planning of such experiments and approval obtained by the full IACUC. The number of animals used, student participation, and demonstration should be considered.[8]
   c. If an animal is used in a procedure involving more than minimal pain or distress it should not be used in additional procedures unless the additional procedure is an integral part of the particular educational protocol and the additional procedure will involve no or minimal pain or distress, or unless the animal is anesthetized throughout the additional procedure and is not allowed to recover. All these procedures should be approved by the institution's full IACUC.

d. All students should receive education and training in the special aspects of anesthesia and surgery for animal experimentation before conducting invasive procedures. Consequently, this material should be included in student curricula.

Reference Notes

The following notes clarify items referred to in the text of the guidelines. They pertain to specific laws or regulations and are included here for your information.

1 This is consistent with the new Public Health Service (PHS) Policy on Humane Care and Use of Laboratory Animals (September 5, 1986) which requires that the IACUC be composed of not fewer than five members and shall include at least one veterinarian, one practicing scientist, one nonscientist, and one individual unaffiliated with the institution. This requirement for five committee members exceeds the provision of the Health Research Extension Act of 1985 (November 20, 1985) for a minimum of three IACUC members. Of these three, there should be at least one veterinarian and at least one individual who has no association with the entity. The new PHS Policy also exceeds the provision of the Improved Standards for Laboratory Animals Act of 1985 (December 23, 1985, which amended the Animal Welfare Act) that the IACUC consist of not fewer than three members possessing sufficient ability to assess animal care, treatment, and practices in experimental research. At least one of these members should be a veterinarian, one should be unaffiliated with the organization and should provide representation of community interests.

2 One of the provisions of both the above Acts is that scientists, animal technicians, and other personnel involved with animal care, treatment, and use have available to them instruction or training in the humane practice of animal maintenance and experimentation, and the concept, availability, and use of research or testing methods that limit the use of animals or limit animal distress.

3 When research projects are conducted under the Good Laboratory Practice Act (GLPA), the investigator should be cognizant of GLPA regulations. Facilities to be used should be precisely defined in order to circumscribe and limit the area that, under GLPA regulations, must meet all the stringent requirements for documentation of conditions as well as identification and care of animals.

4 According to the Improved Standards for Laboratory Animals Act, if procedures to be used will cause pain and/or distress to the animal, a veterinarian should be consulted in the planning of those procedures.

5 Adapted from the Interagency Research Animal Committee's U.S. Government Principles for the Utilization and Care of Vertebrate Animals Used in Testing, Research, and Training (May 20, 1985).

6 Appropriate aseptic technique is required to prevent infection and varies with the type of surgery planned as well as the species being used.

a) According to the PHS *Guide*, aseptic surgery is to be used on most animals, including lagomorphs, that undergo major survival surgery in facilities intended for that purpose. Major survival surgery is defined as any surgical intervention that *penetrates and exposes* a body cavity or has the potential for producing a *permanent handicap* in an animal that is expected to recover.

Aseptic technique for these conditions includes wearing sterile surgical gloves, gowns, caps, and face masks; use of sterile instruments and solutions; and aseptic preparation of the surgical field.

b) Survival surgery in rodents does not require a special facility but should be performed in a "clean" area using sterile instruments, surgical gloves, and aseptic procedures to prevent clinical infections. Similar conditions may also be appropriate for other animals including birds, amphibia, and fish.

[7] Investigators should consider recommendations by the U.S. Department of Agriculture (APHIS, USDA, Subchapter A—Animal Welfare, Jan., 1985) that a source of potable water be provided at least every 6 hours for rabbits, guinea pigs, and hamsters; at least every 12 hours for non-human primates, dogs, cats, and other warm-blooded animals including rats and mice. For those animals for which there is concern about dehydration, and for marine mammals that require drinking water, deprivation should be judged in accordance with the species at a level which would prevent excessive dehydration and would not jeopardize their health.

[8] See note 4.

APPENDIX I. Types of Experiments.

This appendix is designed as a guide to the ethical conduct of animal procedures and as an aid in the selection of humane procedures. It will be of value to scientists designing experiments and to IACUC members in identifying procedures which are considered to be painful and/or distressful in various types of animal research. There should always be careful consideration of the experimental design and the use of proper statistical methods to reduce the number of animals needed. Assignment of particular experiments to any of these types depends on many considerations, such as duration and frequency as well as the nature of the procedure and the species used. Experiments may necessarily involve the use of more than one of these categories. Due consideration should always be given to the necessity for the procedure and the scientific soundness of the project.

Types of experiments	Examples of procedures	Special ethical considerations
A. Experiments on **unanesthetized** animals not expected to have pain and/or distress.	Naturalistic observations of behavior (e.g., in the field, in non-extreme environments); social interactions and non-invasive procedures; EEG, EKG, EMG in trained animals; minimal nutritional modifications; conditioning or training procedures using appetitive motivation; habituation to nonaversive stimuli	Appropriate species, strain; health status and history; duration of procedure; disruption of other coexisting species during field studies.
B. Experiments on **unanesthetized** animals which may be expected to have no more than minimal pain and/or distress.	Procedures that are expected to be no more painful or distressful than administration of a medication or anesthetic agent: experiments involving minimal physical restraint related to physical examination, EEG, EKG, EMG in untrained animals; minimal deprivation; learning or conditioning with aversive stimulation of short duration or non-traumatic magnitude (e.g., mild shock, corneal air puff); injections/collections of blood samples or other body fluids; pain titration; dietary feeding; models of minimal disease; tumor growth not affecting normal function; product safety testing; short and long term studies of effects of chemical or other agents known not to have significant noxious, addicting or intoxicating effects and measured non-invasively; pithing of frogs	As in Type A; duration of procedure; duration of restraint and/or deprivation; pithing should be carried out with the animal immobilized and wherever feasible at a temperature **no greater than** 40°F.

APPENDIX I. *Continued*

Types of experiments	Examples of procedures	Special ethical considerations
C. Experimental use of animals after they have been painlessly killed.	Removal of tissue or organs for study	Method of euthanasia.
D. Acute experiments on **anesthetized** animals which will not be permitted to recover from anesthesia.	Physiological and neurophysiological studies; studies with chemical agents (e.g., CNS stimulants and depressants); non-survival surgical procedures	Mode and effectiveness of anesthesia; if neuromuscular (paralytic) agents are used in combination with anesthesia, appropriate personnel and equipment **must be available** to assure that the animal is ventilated and remains anesthetized.
E. Experimental procedures, including surgical procedures, carried out **under general or regional anesthesia** with intent of recovery. Recovery period with the animal's adaptation to the experimental intervention is considered an integral part of the experimental protocol, and no serious modifications to the animal's behavior, feeding, or ambulatory patterns are anticipated. No more than minimal to moderate pain and/or distress is anticipated which can be altered when necessary by the administration of appropriate analgesic and sedative drugs and proper veterinary care.	Biopsies, catheterizations, implantation of biomedical material, injection of substances to produce inflammatory reaction (e.g. lactic acid, colchicine); abdominal, thoracic or peripheral surgery, (e.g., for implantation of chronic monitoring or support devices and for modification of internal organs); peripheral and central nervous system lesions and implants; pituitary removal; direct treatment, transfer, or sampling of fetuses or embryos in-utero	As in Types A-D; appropriate facilities, aseptic technique and surgical procedures for the species involved; post-operative nursing care which also includes maintenance of body temperature, hydration, and control of pain and infection; duration of procedure; duration of post-operative pain and distress and consideration of use of analgesics; approval of full IACUC is necessary.

APPENDIX I. *Continued*

Types of experiments	Examples of procedures	Special ethical considerations
F. Experimental procedures, including surgical procedures as in E above, but where the study is a progression of the surgical or the preoperative period and where the surgical preparation may be a component of the overall study. The observation and study period after the intervention may produce more than moderate pain, distress, or illness and/or significantly impair the ability of the animal to function in its environment.	Shock; burn and organ transplant and rejection studies; severe trauma; extensive lesions or ablations	As in Type E and additional requirements for analgesia and sedation during remainder of procedure if necessary; approval of full IACUC is required.
G. Experiments on **unanesthetized** animals which may be expected to have more than minimal pain and/or distress and/or where death is anticipated.	Administration of compounds resulting in damage to vital tissue or serious alteration of function; tests of drug efficacy to control severe pain; models of severe disease; severe trauma, injury, or shock; prolonged intense aversive stimulation; extensive tumor growth affecting normal function; studies which do not permit the animal to control the amount of intense painful stimuli it will receive or where drugs will not be administered to modify severe pain or distress; highly invasive studies of major sensory pathways; short and long term studies of the effects of chemical or other agents but known or expected to have significant noxious, addicting, or intoxicating effects	As in Type F and additional requirements for analgesia and sedation during remainder of procedure if necessary; specific justification is required to approve traumatic studies without proper anesthesia, analgesia, or sedation or with the non-surgical use of neuromuscular (paralytic) agents without sedation or anesthesia; such studies must be conducted with special considerations for comfort of the animal and the number of animals used; the investigator must specify the end-point of the experiment as well as alternative situations in which termination of the experiment would be mandatory to avoid prolongation of suffering; such specification may be based on pathological, physiological, or behavioral observation; these procedures must not be carried out without the approval of the full IACUC.

APPENDIX II. Checklist for Design and Review of Proposals

This checklist is intended for use by Institutional Animal Care and Use Committees as an aid in their review of proposed use of animals. This is not an exhaustive checklist, but does include relevant areas across a variety of disciplines. It should be considered in conjunction with the more detailed discussion in the guidelines. Institutions should add review questions pertinent to particular procedures as well as species being used by their investigators. This checklist encompasses the principles and guidelines to which they are appended.

1. Has consideration been given to the use of non-animal methods?
   Is the species selected the "least complex" that is appropriate to answer the question being posed?
   Is the number of animals to be used the minimum that is appropriate to answer the question being posed?
   Can the animal or its organs or tissues be shared with others at any point in the procedure?
   Will surviving animals be offered to other investigators or instructors?
   If yes, do these individuals have approved protocols for such use?
2. Will purpose-bred animals be used?
   If not, will animals be lawfully acquired (consider country of origin and federal, state, and local laws)?
   Is the animal being used a member of a threatened or endangered species?
   Is its use essential?
   Is animal care, use, and housing in accordance with federal, state, and local laws and regulations?
   When were conditions last verified and by whom?
3. Who are the professional personnel responsible for use, care, and handling of animals?
   What are their qualifications?
   What type of expertise should be added, if any?
4. Is the level of anticipated pain or distress essential for the purpose of the experiment?
5. Has consideration been given to the use of non-invasive methods?
   What steps will be taken to minimize pain, distress, or lasting harm?
   Is the anesthesia, analgesia, or other method to reduce pain or distress appropriate and is its duration sufficient for all procedures proposed?
   If reinforcement of animal behavior will be utilized, is it positive?
   If aversive stimulation is proposed, is it minimal with regard to intensity, duration, and frequency within limits of experimental compatibility? How will this be assessed?
   If aversive stimulation is proposed, will the animal be allowed to prevent, escape, or control the intensity of experimental pain?
6. Does the experiment involve multiple procedures on the same animal?
   Are they an integral and necessary part of the experiment?
   If any of these procedures involve pain and/or distress, will additional procedures involve more than minimal pain and/or distress?
   Is the additional procedure one in which the animal is subject throughout to general anesthesia from which it will not be allowed to recover?
7. If surgery is required will appropriate aseptic technique and anesthesia be provided along with appropriate post-operative analgesia and care?
8. If neuromuscular blocking agents are used, will appropriate anesthetics or analgesics be used?
   If yes, are the personnel and equipment adequate to assure that animals will remain anesthetized or will be given analgesics throughout the entire procedure?
   Is the omission of anesthesia or analgesia justifiable and legitimate?
9. If prolonged physical restraint is proposed, is it of the minimal duration and degree essential to the experimental design?

APPENDIX II. *Continued*

10. Will any essential parts of the animal's life support system be altered significantly, e.g., withholding or oversupplying liquids or food or preventing sleep?
    Have short and long term effects been considered, e.g., liquid deprivation for more than 12 hours in birds, or food deprivation that results in a weight loss of greater than 20%?
11. Will the state of the animal be altered by use of analgesics, anesthetics, other pharmacological agents, toxic substances, or surgical procedures?
    If yes, how will the animal be protected from accidental or self-inflicted injury?
    In the case of physiological or psychological dependence on drugs, how will the animal's withdrawal be monitored to minimize the pain and distress of withdrawal?
12. Will animals be euthanized and how?
    Is the method appropriate to the experiment? (see Appendix III)
    Is the procedure in compliance with AVMA recommendations?
    If not, is the proposed procedure essential for the experiment?
    Does the experiment provide for the immediate euthanizing of any animal that will suffer severe acute or chronic pain, distress, or disablement?
    If death is anticipated, has every effort been made to establish a predetermined point at which animals can be euthanized to prevent unnecessary suffering?
13. If animals will be used for instructional purposes, is individual use of animals by students essential?
    Will students be instructed in the handling and care of animals and the effects of the laboratory environment and stressors on the animal and on experimental results?
    Will anesthesia and analgesia for animal use be part of the student's curriculum?
    Have video tapes, films or demonstrations been considered wherever feasible, (e.g., to demonstrate known phenomena, or wherever live animal use is not essential)?
    Does the curriculum include instruction on ethical issues in animal experimentation and the use of alternatives?

Appendix III. Selections From the 1986 Report of the AVMA Panel on Euthanasia†

Preface

In 1984, at the request of the AVMA Council on Research, the Executive Board of the AVMA appointed a Panel on Euthanasia consisting of six veterinarians and one public representative. The purpose of the panel was to review and update the third Panel Report published in 1978[1] Since 1978 the panel has become aware of a need for additional information

Members of the Panel

A. W. Smith, DVM, PhD (Chairman), College of Veterinary Medicine, Oregon State University, Corvallis, OR 97331

K. A. Houpt, VMD, PhD, New York State College of Veterinary Medicine, Cornell University, Ithaca, NY 14853

R. L. Kitchell, DVM, PhD, School of Veterinary Medicine, University of California, Davis, CA 95616

D. F. Kohn, DVM, PhD, University of Texas Medical School at Houston, PO Box 20708, Houston, TX 77225

L. E. McDonald, DVM, PhD, College of Veterinary Medicine, University of Georgia, Athens, GA 30601

Martin Passaglia, Jr., MS, Executive Director, The American Humane Association, 1979–1985

J. C. Thurmon, DVM, MS, College of Veterinary Medicine, University of Illinois, Urbana, IL 61801

E. R. Ames, DVM, PhD, Staff Coordinator, American Veterinary Medical Association, Schaumburg, IL 60196

†Reprinted from the Journal of the American Veterinary Medical Association, Vol. 188, pages 252–268, 1986, with permission of The American Veterinary Medical Association. These selections provide an overview of the AVMA panel report. The reader is strongly encouraged to read the complete report, especially for those agents or methods marked with an asterisk (*) in Table 2.

on some aspects of euthanasia. In this report the panel has expanded the information on research uses of animals wherein euthanasia is required, on public disposal of surplus animals, and on slaughter of food animals. The panel is aware that there are euthanatizing methods and agents not discussed, but has limited this report to those methods and agents supported by reliable information. The report will be targeted primarily to veterinarians, but will be understandable to a broad segment of the general population. Although the interpretation and use of this panel report cannot be limited, the panel would remind all who refer to it that our overriding commitment is to give professional guidance for relieving the pain and suffering of animals.

Introduction

Euthanasia is the act of inducing painless death. Criteria to be considered for a painless death are: rapidly occurring unconsciousness and unconsciousness followed by cardiac or respiratory arrest. The distress experienced by people when observing euthanasia or death in any form is an emotional response dependent on the background of the observer. Kinship of people with higher animals, however distant, serves to transfer the unpleasant reaction to human death to death of animals. Such distress occurs even though the observer experiences no physical pain. This distress may be minimized by perfection of the technique of euthanasia. Although not an adequate criterion, observers may mistakenly relate any movement with consciousness and lack of movement with unconsciousness. Techniques in which animals being euthanatized exhibit little or no movement are the most aesthetically acceptable to most people.

Pain must be defined before criteria for a painless death can be established. Pain is that sensation (perception) that results from nerve impulses reaching the cerebral cortex via specific neural pathways called nociceptive pathways. The term nociceptive is derived from noxious stimuli. Noxious stimuli threaten to, or actually do, destroy tissue. They initiate nerve impulses by acting upon a specific set of receptors, called nociceptors. Nociceptors respond to excessive mechanical, thermal, or chemical energies. Endogenous chemical substances such as hydrogen ions, serotonin, histamine, bradykinin, and prostaglandins as well as electrical currents are capable of generating nerve impulses in nociceptors.

Nerve impulse activity generated by nociceptors is conducted to the spinal cord or the brainstem via nociceptor primary afferent fibers. Within the spinal cord or the brainstem the nerve impulses are transmitted to two sets of neural networks. One set is related to nociceptive reflexes and the second set consists of ascending pathways to the reticular formation, thalamus, and cerebral cortex for sensory processing. The transmission of nociceptive neural activity is highly variable. Under certain conditions, both the nociceptive reflexes and the ascending pathways may be suppressed, as, for example, in deep surgical anesthesia. In another set of conditions, nociceptive reflexes may occur, but the activity in the ascending pathways is suppressed; thus the noxious stimuli are not perceived as pain, as, for example, in light surgical anesthesia. It is incorrect to use the term pain for stimuli, receptors, reflexes, or pathways because the term implies perception whereas all of the above may be active without consequential pain perception.[2–6] Pain is divided into two broad categories: (1) sensory-discriminative, which indicates the site of origin and the energy source giving rise to the pain; and (2) motivational-affective, in which the severity of the stimulus is perceived and the animal's response is determined. Sensory discriminative processing of nociceptive impulses is most likely to be accomplished by brain mechanisms similar to those utilized for processing of other sensory discriminative input that provides the individual with information about the intensity, duration, location, and quality of the stimulus.

Motivational-affective processing involves the ascending reticular formation for behavioral and cortical arousal. It also involves thalamic input into both the forebrain and the limbic system for perceptions such as suffering, fear, anxiety, and depression. The motivational-affective neural networks also have strong inputs into the hypothalamus and the autonomic nervous system for reflex activities of the cardiovascular, pulmonary, and hypophyseal adrenal systems. Responses activated by these systems feed back into the forebrain and enhance the perceptions derived via motivational-effective inputs. Based upon neurosurgical experience in people, it is possible to separate the sensory-discriminative components from motivational-affective components of pain.[3] On an anatomic basis, it would appear that in animals the sensory-discriminative pathways are smaller, compared with that of people, whereas motivational-affective pathways are more numerous and more diverse in animals.[3,7] Some have speculated that animals need, more than human beings, the motivational-affective input to warn them of impending danger.

For pain to be experienced, the cerebral cortex and subcortical structures must be functional. An unconscious animal cannot experience pain because the cerebral cortex is not functioning. If the cerebral cortex is rendered nonfunctional by any means such as hypoxia, depression by drugs, electric shock, or concussion, pain is not experienced.

Stimuli that might evoke pain in a conscious animal may elicit only reflex responses manifested by movement in an unconscious animal. For this reason, nonpurposeful movements of an animal are not reliable indicators of pain perception. Conversely, an animal can experience pain, even though no body movements occur in response to noxious stimuli, if the animal is given muscle-paralyzing agents such as curare, succinylcholine, gallamine, pancuronium, nicotine, or decamethonium. These muscle paralyzing agents do not induce unconsciousness or depress the cerebral cortex or any neural mechanism involved in pain perception.

As with other procedures applied to animals, euthanasia requires some physical control over the animal. The degree of control and kind of restraint needed will be determined by the animal species, breed, size, state of domestication, presence of painful injury or disease, degree of excitement, and method of euthanasia. Suitable control is vital to minimize pain in animals, to assure safety of the person performing euthanasia, and, frequently, to protect other animals and people.

Selection of the most appropriate method of euthanasia in any given situation is dependent on species of the animal involved, available means of animal control, skill of personnel, numbers of animals, economic factors, and other considerations. This report deals primarily with domestic animals, but the same humane considerations should be applied to all species.

Behavioral Considerations

The facial expressions and body postures that indicate various emotional states have been described.[8-10] Behavioral and physiologic responses to noxious stimulation include distress vocalization, struggling, attempts to escape, defensive or redirected aggression, salivation, urination, defecation, evacuation of anal sacs, pupillary dilatation, tachycardia, sweating, and reflex skeletal muscle contractions causing shivering, tremors, or other muscular spasms. Some of these responses can occur in unconscious as well as conscious animals. Fear can cause immobility or freezing in certain species, particularly rabbits and chickens. This immobility response should not be interpreted as unconsciousness when the animal is, in fact, conscious.[10]

In very young animals, autonomic and reflexive reactions are evident, although overt behavioral reactions may differ from those of adults.

The need to minimize fear and apprehension must be considered in determining the method of euthanasia. Distress vocalizations, fearful behavior, and release of certain odors or pheromones by a frightened animal may cause anxiety and apprehension in others. Therefore, whenever possible, animals should not observe euthanasia of others, especially of their own species. This is particularly important when vocalization or release of pheromones may occur during induction of unconsciousness. Gentle restraint, preferably in a familiar environment, careful handling, and talking during euthanasia often have a calming effect on companion animals. However, some of these methods may not be operative with wild animals or animals that are injured or diseased. Where capture or restraint may cause pain, injury, or anxiety to the animal or danger to the operator, the use of tranquilizing or immobilizing drugs may be necessary.

The ethical and psychologic issues involved with euthanasia of diseased, injured, or unwanted animals is within the purview of this discussion. Moral and ethical imperatives associated with individual or mass animal euthanasia should be consistent with acceptable humane practice. In all circumstances, choice of method should be selected and employed with the highest ethical standards and social conscience. Many issues involving diseased, injured, or unwanted animals and the need for euthanasia have been addressed in conferences cosponsored by AVMA.[14,15]

Distress may occur among personnel directly involved in performing repetitive euthanasia of diseased, injured, or unwanted animals. At the point of terminating the life of an animal, we should be prepared not only to treat the animal but also to consider the people attached to the animal.[16] Constant exposure to or participation in euthanasia procedures can cause a psychologic state characterized by a strong sense of work dissatisfaction or alienation, which may be expressed by absenteeism, belligerence, or careless and callous handling of the animals.[17] This is one of the principal reasons for turnover of employees directly involved with repeated performance of animal euthanasia. This should be recognized as a bona fide personnel problem related to animal euthanasia, and management measures should be instituted to decrease or eliminate the potential for this condition.

Modes of Action of Euthanatizing Agents

Euthanatizing agents terminate life by three basic mechanisms: (1) hypoxemia, direct or indirect; (2) direct depression of neurons vital for life function; and (3) physical damage to brain tissue (Table 1).

Agents that produce death by direct or indirect hypoxia can act at various sites and can cause different times of onset of unconsciousness. With some agents, unconsciousness may occur prior to cessation of motor activity (muscle movement). Thus, even though animals demonstrate muscular contractions, they are not perceiving pain. The uninformed observer may find this difficult to accept.

Conversely, muscle relaxants induce a flaccid muscular paralysis and the animal remains conscious until death eventually occurs as a result of hypoxemia and hypercapnea. While the animal appears relaxed, it is actually in a state of panic and can feel pain. Outwardly this would appear to be an ideal form of euthanasia, but it is not. Agents that do not induce rapid loss of consciousness prior to death include: curare, succinylcholine, gallamine, nicotine, magnesium or potassium salts, pancuronium, decamethonium, and strychnine. *Use of any of these agents alone for euthanasia is absolutely condemned.*

The second group of euthanatizing agents depresses nerve cells of the brain first, blocking apprehension and pain perception, followed by unconsciousness and death.

Some of these agents "release" muscle control during the first stage of anesthesia, resulting in a so-called "excitement or delirium phase," during which there may be vocalization and some muscle contraction.[18] Although these responses appear to be purposeful, they are not. Death is due to hypoxemia and direct depression of respiratory centers.

With the third group, physical damage to the brain, concussion, or direct electrical flow[19] through the brain produces instant unconsciousness. Exaggerated muscular activity may follow unconsciousness. When electrocution is properly performed, loss of motor function occurs concomitantly with loss of consciousness; with the other methods in this group, muscular contraction may occur. Although this may be unpleasant to the observer, the animal is not suffering.

Electroencephalograms and other physiologic measurements can be employed to measure responses to euthanatizing agents.[1]

Criteria for Judging Methods of Euthanasia

Several criteria were used in evaluating methods of euthanasia: (1) ability to produce death without causing pain; (2) time required to produce loss of consciousness; (3) time required to produce death; (4) reliability; (5) safety of personnel; (6) potential for minimizing undesirable psychologic stress on the animal; (7) nonreversibility; (8) compatibility with requirement and purpose; (9) emotional effect upon observers or operators; (10) economic feasibility; (11) compatibility with histopathologic evaluation; and (12) drug availability and abuse potential (Table 2).

Inhalant Agents

Occupational exposure to inhalation anesthetics constitutes a human health hazard. An increased incidence of spontaneous abortion and congenital abnormalities results from exposure to trace amounts of inhalation anesthetic agents.[20] Human exposure levels for volatile liquid anesthetics (ether, halothane, methoxyflurane, ethrane, and isoflurane) should be less than 2 ppm, and less than 25 ppm for nitrous oxide.[21] While there are no controlled studies proving that such levels of anesthetics are "safe," these concentrations were established because they were shown to be attainable under hospital conditions. Effective procedures must be employed to protect personnel from anesthetic vapors.

Noninhalant Pharmacologic Agents

Noninhalant agents that can be used for euthanasia are widely diverse in chemical composition.[64,65] Although death can be induced by administering these drugs via many routes (intravenous, intracardiac, intraperitoneal, intrathecal, intramuscular, intrathoracic, subcutaneous, and oral), intravenous administration is preferred because the effect is most rapid and reliable. Intrapulmonic injection should be avoided.

TABLE 1—Mode of action of euthanatizing agents

	Site of action	Classification	Comments
	HYPOXIC AGENTS		
Carbon monoxide	Carbon monoxide combines with hemoglobin of RBC, preventing combination with oxygen	Histotoxic hypoxia	Unconsciousness occurs rapidly; motor activity persists after unconsciousness
Hydrogen cyanide	Depression of oxygen transport at tissue level; no O_2 available in tissues	Histotoxic hypoxia	Unconsciousness occurs rapidly; motor activity persists after unconsciousness
Curariform drugs; curare, succinylcholine, gallamine decamethonium	Paralysis of respiratory muscles; oxygen not available to blood	Hypoxic hypoxemia and hypercarbia	Unconsciousness develops slowly, preceded by anxiety and fear; no motor activity
Rapid decompression	Reduced partial pressure of oxygen available to blood	Hypoxic hypoxemia	Unconsciousness occurs rapidly; motor activity persists after unconsciousness
Nitrogen inhalation	Reduced partial pressure of oxygen available to blood	Hypoxic hypoxemia	Unconsciousness occurs rapidly; motor activity persists after unconsciousness
Electrocution when current does not pass through brain	Spastic paralysis of respiratory muscles and ventricular fibrillation; oxygen not available to blood in lungs	Hypoxic hypoxemia	Unconsciousness develops slowly, occurring after violent muscle spasms
	DIRECT NEURON DEPRESSING AGENTS		
Anesthetic gases; ether, chloroform, methoxyflurane, halothane, nitrous oxide, and enflurane	Direct depression of cerebral cortex and subcortical structures and vital centers	Hypoxemia due to depression of vital centers	Unconsciousness occurs first; no anxiety or pain; possible involuntary motor activity after unconsciousness; no motor activity after brief period
Carbon dioxide	Direct depression of cerebral cortex, subcortical structures and vital centers; direct depression of heart muscle	Hypoxemia due to depression of vital centers	Unconsciousness occurs first; no anxiety or pain; possible involuntary motor activity after unconsciousness; no motor activity after brief period
Barbituric acid derivatives	Direct depression of cerebral cortex, subcortical structures and vital centers; direct depression of heart muscle	Ultimate cause of death is hypoxemia due to depression of vital centers	Unconsciousness reached rapidly; no anxiety; no excitement period; no motor activity; best to administer by intravenous or intracardiac administration
Chloral hydrate and chloral hydrate combinations	Direct depression of cerebral cortex, subcortical structures and vital centers; direct depression of heart muscle	Ultimate cause of death is hypoxemia due to depression of vital centers	Transient anxiety; unconsciousness occurs rapidly; no motor activity
T-61	Direct depression of cerebral cortex, subcortical structures and vital centers; direct depression of heart muscle	Hypoxemia due to depression of vital centers	Transient anxiety and struggling may occur before unconsciousness when given too rapidly; tissue damage may occur. Must be given intravenously at recommended dosage and rates

		Physical Agents	
Gunshot or captive bolt into brain or stunning	Direct concussion of brain tissue	Hypoxemia due to depression of vital centers	Instant unconsciousness; motor activity may occur after unconsciousness
Cervical dislocation	Direct depression of brain	Hypoxemia due to disruption of vital centers	Violent muscle contractions can occur after cervical dislocation
Decapitation	Direct depression of brain	Hypoxemia due to disruption of vital centers	Violent muscle contractions occur subsequent to decapitation
Exsanguination	Direct depression of brain	Hypoxemia	If this method is preceded by and occurs during unconsciousness, there should be no struggling or muscle contraction
Decompression	Direct depression of brain	Hypoxemia	There are numerous disadvantages associated with mechanical problems as well as painful physiologic problems associated with expanding gases
Electrocution through brain	Direct depression of brain	Hypoxemia	Violent muscle contractions occur at same time as unconsciousness
Microwave irradiation	Direct inactivation of brain enzymes by rapid temperature increase of brain	Brain enzyme inactivation	Currently used only in rodents
Rapid freezing	Direct depression of brain	Rapid or near instantaneous cessation of metabolism	Approved only in specialized, well-controlled cases, in small laboratory animals
Air embolism	Direct depression of brain	Hypoxemia due to circulatory collapse	No motor activity occurs when preceded by anesthetics

TABLE 2 — Characteristics of agents and methods of euthanasia

Agent	Safety for personnel	Ease of performance	Rapidity	Economic considerations
		INHALANTS		
Ether	Flammable and explosive	Easily performed with closed chamber or container	Slow onset of anesthesia	Relatively inexpensive
Halothane*	Nonflammable and nonexplosive; chronic exposure of animals or personnel to vapor may be harmful	Easily performed with closed container; can be administered to large animals by means of a mask	Rapid onset of anesthesia	Expensive
Methoxyflurane†	Nonflammable and nonexplosive under normal environmental conditions; chronic exposure of animals or personnel to vapor may be harmful	Slow with closed container; may be difficult to obtain necessary vapor concentration; can be administered to large animals by means of a mask	Slow onset of anesthesia	Expensive
Enflurane‡	Nonflammable and nonexplosive; chronic exposure of animals or personnel to vapors may be harmful	Easily performed with closed container or chamber; can be administered to large animals by means of a mask	Rapid onset of anesthesia	Expensive
Isoflurane‡	Nonflammable and nonexplosive; chronic exposure of animals or personnel to vapors may be harmful	Easily performed with closed container or chamber; can be administered to large animals by face mask	High volatility and potency; rapid onset of anesthesia	Very expensive
Nitrous oxide	Nonflammable and nonexplosive, but will support combustion; chronic exposure of animals or personnel to gas may be harmful	Easily performed with closed container or chamber	Rapid onset in 100% concentration	Relatively expensive
Chloroform	Nonflammable and nonexplosive; chronic exposure of animals or personnel is dangerous because of potential liver or kidney damage and carcinogenicity	Easily performed with closed container; can be administered to large animals by means of a mask	Rapid onset of anesthesia	Inexpensive
N_2	Safe if used with ventilation	Use closed chamber with rapid filling	Rapid	Inexpensive
Hydrogen cyanide	Extremely hazardous to personnel	Airtight chamber required unless used under field conditions	Rapid	Inexpensive
Carbon monoxide	Extremely hazardous; toxic and difficult to detect	Requires appropriately operated equipment for gas production	Moderate onset time, but insidious, so animal is unaware of onset	Inexpensive when proper equipment is in place

TABLE 2 – *Continued*

Tissue changes	Efficacy	Species suitability	Remarks
	INHALANTS		
Slight changes may occur in parenchymatous organs	Highly effective provided that subject is sufficiently exposed	Suitable for cats, young dogs, birds, rodents, and other small species	Acceptable but dangerous
		Administration to larger animals requires specialized equipment and training onset is slow; vapors are irritating	Not acceptable
May occur in parenchymatous organs	Highly effective provided that subject is sufficiently exposed	Suitable for cats and young dogs, birds, rodents, and other small species	Acceptable
May occur in parenchymatous organs	Highly effective provided that subject is sufficiently exposed	Suitable for cats, young dogs, birds, rodents, and other small species	Acceptable
		Administration to large animals requires specialized equipment and training; slow onset	Not acceptable
May occur in parenchymatous organs, particularly kidneys	Highly effective provided that subject is sufficiently exposed; deep anesthesia may be accompanied by motor activity (twitching)	Suitable for cats, young dogs, birds, and other small species	Acceptable, but not recommended because of motor activity in deep plane of anesthesia
		Useful in large animals in emergency situations	Not acceptable except in emergencies
May occur in parenchymatous organs	Highly effective provided that subject is sufficiently exposed; induction does not appear to be stressful	Suitable for all small animals including birds and rodents	Acceptable
		Useful in large animals in emergency situations	Acceptable only in emergency situations
Hypoxic lesions may occur	Highly effective provided that subject is sufficiently exposed	Suitable for cats, small dogs, birds, rodents, and other small species	Acceptable
		Not recommended alone for larger animals	Use in larger animals requires supplementation with other agents
Extensive changes may occur in parenchymatous organs	Highly effective provided that subject is sufficiently exposed	Suitable for cats, young dogs, birds, rodents, and other small species	Acceptable only in controlled conditions (*see* text)
		Useful in large animals in emergency situations	Acceptable only in emergency situations
Changes associated with hypoxemia may occur	Effective except in young and neonates	Suitable for most small species, including mink	An effective agent, but other methods preferable; not acceptable in most animals less than 4 months old
Changes consistent with tissue hypoxia may occur	Effective and irreversible	Dens of foxes, badgers, rabbits, and rodents	Because of extreme danger to operators, other methods are preferred
Changes associated with hypoxemia may occur	Effective	Most small species including dogs, cats, rodents, mink, and chinchillas	Acceptable only when properly designed and operated equipment is used

TABLE 2 – *Continued*

Agent	Safety for personnel	Ease of performance	Rapidity	Economic considerations
Carbon dioxide	Minimal hazard	Used in closed container	Moderately rapid	Inexpensive
		NONINHALANTS		
Barbiturates	Safe except human abuse potential; DEA-controlled substance	Animal must be restrained; personnel must be skilled in IV injection	Rapid onset of anesthesia	Relatively inexpensive
Secobarbital/dibucaine	Safe	Animal must be restrained; personnel must be skilled in IV injection	Rapid onset of anesthesia	Expensive
T-61	Safe	Animal must be restrained; personnel must be skilled in IV injection	Rapid onset of anesthesia	Expensive
Mixture of chloral hydrate, $MgSO_4$, pentobarbital	Safe except human abuse potential; DEA-controlled substance	Animal must be restrained; personnel must be skilled in IV injection	Rapid onset of anesthesia	Relatively inexpensive
Strychnine/nicotine/curariform drugs/$MgSO_4$/KCl	...	...	...	...
		PHYSICAL METHODS		
Captive bolt, gunshot and stunning	Safe, but some concern for mechanical injury; gunshot can be especially dangerous	Requires skilled individuals; however, skills easily developed	Rapid	Inexpensive
Cervical dislocation	Safe	Requires training and skill	Moderately rapid	Inexpensive
Decapitation	Some concern for mechanical injury	Easily performed with minimal training	Moderately rapid; can have consciousness for 13 to 14 seconds	Inexpensive
Exsanguination	Safe	Easily performed with minimal training	Moderately rapid	Inexpensive
Decompression	Safe	Requires special equipment that must be operated and maintained by skilled personnel	Relatively slow	Inexpensive
Electrocution	Hazardous to personnel	Not easily performed in all instances	Can be rapid	Inexpensive
Microwave	Safe	Requires training and highly specialized equipment	Very rapid	Equipment is expensive
Rapid freezing	Safe	Easily performed	Rapid	Inexpensive
Air embolism	Safe	Easily performed	Moderately rapid	Inexpensive

*Fluothane, Ayerst Laboratories, New York, NY: Halocarbon Laboratories, Hackensack, NJ.
†Metofane, Pitman-Moore, Inc, Washington Crossing, NJ; Penthrane, Abbott Laboratories, North Chicago, Ill.
‡Ethrane and Isoflurane, Anaquest, Division of BOC, Inc, Madison, Wis.

TABLE 2 — *Continued*

Tissue changes	Efficacy	Species suitability	Remarks
Changes associated with hypoxemia may occur	Effective	Small laboratory animals, birds, cats and small dogs	Acceptable, but time required may be prolonged in immature and neonate animals
		NONINHALANTS	
Drug residues	Highly effective when appropriately administered	All species	Acceptable IV (*see* text)
Drug residues	Highly effective when appropriately administered	Dogs and cats and some laboratory animals	Acceptable IV
Drug residues	Highly effective when appropriately administered	Dogs and cats and some laboratory animals	Acceptable IV (*see* text)
Drug residues	Highly effective when appropriately administered	Large animals, ie, horses and cattle	Acceptable IV
...	...	...	Absolutely unacceptable when used alone (*see* text)
		PHYSICAL METHODS	
Trauma of brain tissues; others unchanged	Highly effective	Usually applied in larger agricultural animals, but can be used for rabbits and guinea pigs	Acceptable (*see* text for limitations)
Primarily useful if chemical residue-free tissues are needed	Irreversible	Suitable only in chickens, laboratory mice, and rats less than 200 g or rabbits less than 1 kg	Acceptable with prior sedation or light anesthesia (*see* text)
Primarily useful if chemical residue-free tissues are needed	Irreversible	Suitable for rodents and small rabbits	Acceptable with prior sedation or light anesthesia (*see* text)
Minimal	Irreversible	Recommended to ensure death following stunning by captive bolt, or gunshot in large domestic species or rabbits	Acceptable when preceded by other methods that relieve anxiety, consciousness
Trauma can occur to tissues when air is trapped in middle ear and gut	Effective if equipment is properly operated and maintained	Primarily for dogs and smaller animals	Not recommended because of numerous disadvantages (*see* text)
Petechial hemorrhages can occur	Highly effective if properly performed	Used primarily in farm animals	Acceptable; however, disadvantages far outweigh advantages in most applications (*see* text)
Primarily used for brain tissue studies	Highly effective	Used in mice, rats, or animals of similar size	Acceptable only in small rodents (*see* text)
Arrest enzymatic reactions in tissue, usually brain	Effective	Used in small rodents	Acceptable (*see* text)
Hypoxemic lesions may occur	Effective if closely monitored	Used primarily in rabbits and other small species	Acceptable only in fully anesthetized animals

Oral, rectal, and intraperitoneal routes of administration of drugs for euthanasia are inadvisable because of prolonged onset of action, wide range in lethal doses, and potential irritation of tissues. An hour or more may elapse from time of administration to death. Some drugs, such as chloral hydrate and T-61, when given intraperitoneally are irritating and cause abdominal pain. Others may produce tissue changes, depending on dose and route of administration.[25,63]

Because crying and struggling may follow improper intracardiac injection, this route of administration is objectionable. Skill is required to penetrate the heart of an animal with one thrust of a hypodermic needle, especially if the animal is not easily restrained. Intracardiac injection of drugs is not recommended for euthanasia, except in depressed, anesthetized, or comatose animals. Intrathecal use of drugs in unanesthetized animals is not recommended because puncture of the cisterna magna without causing pain and struggling is not possible.

If the animal to be euthanatized is excitable or vicious, use of analgesic, tranquilizers,[44] narcotics, ketamine, xylazine, or other depressants is recommended before administration of the euthanatizing agent.

Other Parenteral Preparations

A number of injectable agents are currently available for immobilizing domestic and feral animals. Drugs such as ketamine HCl and xylazine are used for immobilization, analgesia, and anesthesia. The effect occurs within minutes following intramuscular injection, making these drugs useful for restraint (ie, phase I of euthanasia). Currently these agents are not included as controlled substances by federal drug regulations. Narcotics also may be used for sedation and analgesia, but detailed records must be kept, and their use properly supervised by a licensed professional registered with the US Drug Enforcement Agency. Although it is possible to cause death with high doses of these drugs, such use would be impractical and may produce convulsions before death.

Physical Methods

Physical methods include captive bolt pistol, gunshot, stunning, cervical dislocation, decapitation, exsanguination, decompression, electrocution, microwave irradiation, rapid freezing, and air embolism. It is a requirement that any method of euthanasia be performed by knowledgeable, well-trained individuals. Because trauma often is associated with physical methods for inducing unconsciousness, this requirement is particularly vital. Physical methods are often aesthetically displeasing; however, when properly performed, unconsciousness is rapid, without distress or pain to the animal. In some applications the amount of apprehension of the animals will be less than if chemical means had been used.

Precautions Concerning Less Common Species

When euthanasia of poikilothermic (cold blooded) animals and aquatic animals is performed, the differences in their metabolism, respiration, and tolerance to cerebral hypoxemia may preclude methods that would be acceptable in terrestrial mammals.

Postface

This panel report summarizes contemporary scientific knowledge on euthanasia in animals and calls attention to the lack of scientific reports assessing pain and discomfort in animals undergoing euthanasia. Many reports on various methods of euthanasia are either anecdotal or testimonial narratives and are therefore not cited in this report. The panel unanimously endorses the need for well-designed experiments to more fully determine the extent to which each procedure meets the criteria used for judging the methods of euthanasia.

The Panel on Euthanasia is fully committed to the concept that whenever it becomes necessary to kill any animal for any reason whatsoever, death should be induced as painlessly as possible. It has been our charge to develop workable guidelines for addressing this need, and it is our sincere desire that these guidelines be conscientiously used by all who exercise stewardship over animals on earth.

References

[1] McDonald LE, Booth, NH, Lumb WV, et al. Report of the AVMA panel on euthanasia. *J Am Vet Med Assoc* 1978;173:59–72.

[2] Breazile JE, Kitchell RL. Euthanasia for laboratory animals. *Fed Proc* 1969;28:1577–1579.

[3] Kitchell RL, Erickson HH, Carstens E, et al, eds. *Animal pain: perception and alleviation.* Bethesda, Md, Am Physiol Soc, 1983.

[4] Kitchell RL, Johnson RD. Assessment of pain in animals. In: Moberg GP, ed. *Animal stress* Am Physiol Soc, 1985;113–140.

[5] Willis WD. *The pain system. The neural basis of nociceptive transmission in the mammalian nervous sytem.* Basel, Switzerland: S Karger, 1985;346.

[6] Zimmerman M. Neurobiological concepts of pain, its assessment and therapy. In: Bromm B, ed. *Pain measurement in man. Neurophysiological correlates of pain.* Amsterdam: Elsevier Publishing Co, 1984;15–35.

[7] Dennis SG, Melzack, R. Perspectives on phylogenetic evolution of pain expression. In: Kitchell RL, Erickson HH, Carstens E, et al, eds. *Animal pain: perception and alleviation.* Bethesda, Md, Am Physiol Soc, 1983.

[8] Beaver B. *Veterinary aspects of feline behavior.* St Louis: CV Mosby Co, 1980.

[9] Hafez ESE. *The behavior of domestic animals.* 3rd ed. Baltimore: The Williams & Wilkins Co, 1975.

[10] Houpt KA, Wolski TR. *Domestic animal behavior for veterinarians and animal scientists.* Ames, Iowa: Iowa State University Press, 1982;356.

[14] *Proceedings of the national conference on dog and cat control.* Denver: Am Vet Med Assoc, 1976.
[15] *Proceedings of the national conference on the ecology of the surplus dog and cat problem.* Chicago, Ill: Am Vet Med Assoc, 1974.
[16] Bustad LK. An educator's approach to euthanasia. *Lab Anim* 1982;11:37–41.
[17] Wolfle TL. Laboratory animal technicians: their role in stress reduction and human-companion animal bonding. *Vet Clin North Am [Small Anim Pract]* 1985;15:441–454.
[18] Lumb WV, Jones E. *Veterinary anesthesia.* Philadelphia: Lea & Febiger, 1985.
[19] Warrington R. Electrical stunning, a review of literature. *Vet Bull* 1974;44:617–628.
[20] *Occupational exposure to waste anesthetic gases and vapors.* Washington, DC: Department of Health, Education, and Welfare (National Institute for Occupational Safety and Health) No. 77–140, 1977.
[21] Lecky JH, ed: Waste anesthetic gases in operating room air: a suggested program to reduce personnel exposure. *Special report.* Park Ridge, Ill: The American Society of Anesthesiologists, 1983.
[25] Sawyer DC: Comparative effects of halothane. *Gaines Dog Res Prog* 1976, 2–3.
[44] Dallaire A, Chalifoux A. Premedication of dogs with acepromazine or pentazocine before euthanasia with carbon monoxide. *Can J Comp Med* 1985;49:171–178.
[63] Feldman DB, Gupta BN. Histopathologic changes in laboratory animals resulting from various methods of euthanasia. *Lab Anim Sci* 1976;26:218–221.
[64] Hatch RC. Euthanatizing agents. In: Booth NH, McDonald LE, eds. *Veterinary pharmacology and therapeutics.* 5th ed. Ames, Iowa: Iowa State University Press, 1982;1059–1064.
[65] Lumb WV. Euthanasia by noninhalant pharmacologic agents. *J Am Vet Med Assoc* 1974;165:851–852.

APPENDIX D

Chronology of Significant Events, 1966–1989

1966 The Laboratory Animal Welfare Act (PL 89–544) is enacted.

1967 The American Association for Laboratory Animal Science (AALAS) is formed as a national organization from the former Animal Care Panel, which had existed since 1950.

1970 The Animal Welfare Act (PL 91–579) is enacted. This is an amended version of the Laboratory Animal Welfare Act of 1966.

1973 Representative Les Alpin leads a letter-writing campaign to protest the use of beagles to test poisonous gases by the army and the air force. As a result, funding for this research is curtailed the following year.

1974 Australian philosopher Peter Singer teaches a course on animal rights at New York University. One of the students in that class is Henry Spira, who later becomes an effective campaigner against the Draize and LD-50 toxicity tests.

1975 Publication of *Animal Liberation* by Peter Singer (New York: New York Review/Random House) gives new impetus to the animal rights movement.

1976 The Animal Welfare Act amendments of 1976 (PL 94–279) are enacted.

Demonstrations in front of the American Museum of Natural History in New York are conducted for several months by animal rights groups protesting the museum's research on cats. The protests gain national publicity, prompting Congressional and NIH investigations. No wrongdoing is found, but the research is terminated the following year.

1978 NIH revises its *Guide for the care and use of animals* for the first time since its original publication in 1963.

Good Laboratory Practice regulations (GLPs) are approved by the FDA and published in the *Federal Register* in December, to go into effect June 20, 1979.

1979 Scientists' Center for Animal Welfare (SCAW) is founded.
New York University Medical Center is raided by members of the Animal Liberation Front. Several research animals are stolen.
Henry Spira, animal rights activist, launches a campaign against Draize eye irritancy test.

1980 Revlon announces a $750,000 grant to Rockefeller University to develop alternatives to the Draize eye irritancy test.
The Federation of American Societies for Experimental Biology (FASEB) adopts "Guiding Principles in the Care and Use of Animals" (the first revision of FASEB's original policy statement in 1913).

1981 The Cosmetic, Toiletry and Fragrance Association (CTFA) donates $1 million to Johns Hopkins University to establish a center to develop alternatives to animal testing.
The FDA drops its requirement for the Draize test of product safety and announces that it will accept results from properly validated alternatives.
Edward Taub, researcher at a Maryland laboratory, is indicted for cruelty to monkeys he was using in neurological experiments. Taub is convicted on six counts of failure to provide proper veterinary care, and his NIH grant is temporarily rescinded. This is the first time NIH suspended a grant because of questions regarding the treatment of laboratory animals.

1982 The Department of Defense bans the use of cats, dogs, and nonhuman primates in nuclear weapons research.
NIH concludes its investigation of the Taub case, ruling that Taub violated its guidelines for the care and use of research animals. Taub's NIH grant is permanently revoked.

1983 Twelve dogs being used in a heart pacemaker study are stolen from the University of California-Harbor Medical Center near Los Angeles.
Military wound research using animals is halted by Defense Secretary Caspar W. Weinberger (but the following year the ban is limited to cats and dogs).
Taub's conviction is reversed by the Maryland Court of Appeals, but his NIH grant is not reinstated.

1984 PHS publishes proposed revisions to the Public Health Service Policy on Humane Care and Use of Animals by Awardee Institutions.
A break-in at a University of Pennsylvania head trauma laboratory is carried out by members of the Animal Liberation Front (ALF). Stolen videotapes of baboon experiments are widely shown on national television, prompting an investigation by NIH.

The Federation of American Societies for Experimental Biology (FASEB) adopts a revision of its 1980 policy statement concerning the care and use of reseach animals.
EPA announces new guidelines discouraging the use of standard LD-50 tests using 200 animals.
In a second break-in at the University of Pennsylvania by members of the Animal Liberation Front, animals are stolen from the School of Veterinary Medicine.
More than 100 animals are stolen from the City of Hope Medical Center in Duarte, California, by Animal Liberation Front members. NIH investigates charges of animal abuse.

1985 The American Psychological Association adopts new guidelines for animal experimentation (a revision of its 1979 guidelines).
Four hundred sixty-seven animals are stolen from the animal facility at the University of California (Riverside); an NIH investigation follows.
PHS releases its revised Policy on Humane Care and Use of Laboratory Animals by Awardee Institutions.
HHS Secretary Margaret Heckler suspends funding for the head trauma research at the University of Pennsylvania in July, and 2 months later the university suspends the research program indefinitely.
NIH suspends funding to the City of Hope Medical Center, and the Department of Agriculture fines the institution for violations of the Animal Welfare Act.
The University of Pennsylvania announces its intent to establish the nation's first professorship in humane ethics and animal welfare, at the School of Veterinary Medicine. (The chair, to be named for the donor, the late Marie A. Moore, had not yet been filled as of early 1989.)
The PHS Health Research Extension Act (PL 99–158) is enacted.
The Animal Welfare Act amendments of 1985 (PL 99–198) are enacted.
NIH publishes a revision of its *Guide for the care and use of animals* (last revised 1978) to reflect changes mandated by the Health Research Extension Act and revised PHS Policy.

1986 NIH suspends funding for research using vertebrates other than rodents at the Health Sciences Division of Columbia University. This is NIH's first enforcement action as a result of the new PHS policy.
NIH decides to take no action against the University of California (Riverside) in allegations brought by animal rights groups.
The Office of Technology Assessment (OTA) publishes its long-awaited report on laboratory animal use, *Alternatives to animal use in research, testing, and education.*

PHS publishes an amended version of its Policy on Humane Care and Use of Laboratory Animals to incorporate new mandates of the Health Research Extension Act of 1985.

1987 The Supreme Court upholds lower court rulings that animal rights groups have no legal standing to sue to force the removal of Taub's surviving monkeys from NIH custody to a sanctuary in Texas. Later in the year, a new furor erupts when the director of the NIH research center where the monkeys are being held announces that 8 of the 14 monkeys should be euthanized because of their poor condition. Animal rights organizations protest and NIH promises not to kill the monkeys until "further assessments" can be made.

Twenty-eight cats are stolen from the Beltsville Agricultural Research Center in Maryland; a group called the Band of Mercy claims responsibility. According to NIH, there have been 26 such raids since 1981.

1988 NIH rescinds funding for a project to study head trauma in cats at the University of Cincinnatti College of Medicine. The NIH move was prompted by protests from animal protectionists, including the Physicians Committee for Responsible Medicine.

Michiko Okamoto, a researcher at Cornell Medical College who was studying barbiturate dependence in cats, returns a $600,000 grant to the National Institute on Drug Abuse (NIDA) after her experiments are criticized by members of Trans-Species Unlimited, an animal rights group.

The National Academy of Sciences-National Research Council releases its report, *The use of laboratory animals in biomedical and behavioral research.* One committee member refuses to sign the report because it does not "face the widespread, ingrained problem of unnecessary suffering."

The Hastings Center establishes a committee on the ethics of animal experimentation. The committee, under the direction of philosopher Strachan Donnelley, is expected to issue a set of ethical guidelines by the end of 1989.

The first patent for a higher form of life is issued to Harvard University. The patent, for a genetically altered breed of mouse that has a tendency to develop breast cancer, becomes the focus of new conflicts over the morality of patenting living creatures and of creating transgenic animals.

1989 The National Research Council-Institute of Laboratory Animal Resources forms a committee to prepare a handbook on the recognition and alleviation of pain and distress in laboratory animals. An 18-month study period is allotted for the project.

Appendix E

Journals in Which Research Animal-Welfare-Related Items Were Found

The 659 journal items upon which the quantitative analysis was based were found in the following journals:

Journals in which research animal-welfare-related items were found, listed by category.

	Number of relevant items, 1966–1985
Biomedical (54 journals)	
Advances in Nursing Science	2
Alabama Journal of Medical Science	1
American Journal of Kidney Disease	1
American Journal of Public Health	4
American Journal of Surgery	1
American Pharmacy	1
Annual Review of Pharmacology and Toxicology	1
Archives of Internal Medicine	1
Bioessays	1
Biologist	1
BioScience	18
Bulletin of the American College of Surgeons	2
Bulletin of the History of Medicine	4
California Medicine	1
Cancer Investigation	2
Circulation	2
Circulation Research	1
Circulatory Shock	1
Clinical Research Practices and Drug Regulatory Affairs	1
Dental Students Magazine	1
Environmental Health Perspectives	1
Federation Proceedings	4
Fundamentals of Applied Toxicology	4
Harvard Medical Alumni Bulletin	1
Hospital Progress	1
Hospitals	1
Illinois Dental Journal	1
Illinois Research	1

Journals in which research animal-welfare-related items were found, listed by category.

	Number of relevant items, 1966–1985
International Journal of Dermatology	1
Journal of Applied Physiology, Respiratory, Environmental, and Exercise Physiology	1
Journal of Biocommunication	1
Journal of Clinical Pharmacology	1
Journal of Legal Medicine	1
Journal of Medical Education	2
Journal of Oral Surgery	2
Journal of Pharmaceutical Sciences	1
Journal of Surgical Research	1
Journal of the American Medical Association	12
Journal of the History of Medicine and Allied Sciences	2
Mayo Clinic Proceedings	2
MD	3
Modern Medicine	1
New England Journal of Medicine	17
New York State Journal of Medicine	1
Pavlovian Journal of Biological Science	2
Perspectives in Biology and Medicine	3
Pharmaceutical Technology	3
Pharos	1
Physiologist	64
Proceedings of the Institute of Medicine, Chicago	2
Psychopharmacology Bulletin	4
Surgery	1
Transactions of the American Society for Artificial Internal Organs	1
Transactions and Studies of the College of Physicians of Philadelphia	1
Total	192
Veterinary and animal science (11 journals)	
Advances in Veterinary Science and Comparative Medicine	1
Animal Regulation Studies	5
Applied Animal Ethology	1
California Veterinarian	9
Cornell Veterinarian	1
Journal of Animal Science	7
Journal of the American Veterinary Medical Association	298
Journal of Veterinary Medical Education	1
Laboratory Animal Science	11
Modern Veterinary Practice	1
Proceedings. Annual Meeting of the U.S. Animal Health Association	2
Total	337
Psychology and behavioral sciences (6 journals)	
American Journal of Psychiatry	1
American Psychologist	16
Contemporary Psychology	2

Journals in which research animal-welfare-related items were found, listed by category.

	Number of relevant items, 1966–1985
Journal of Comparative and Physiological Psychology	1
Journal of the Experimental Analysis of Behavior	1
Psychological Record	4
Total	25
General science (8 journals)	
Hastings Center Report	6
Isis	1
Science	89
Science Digest	1
Science Teacher	2
Science, Technology, and Human Values	4
The Sciences	1
Scientific American	1
Total	105

Index

MIX
Papier aus verantwortungsvollen Quellen
Paper from responsible sources
FSC® C105338

If you have any concerns about our products,
you can contact us on
ProductSafety@springernature.com

In case Publisher is established outside the EU,
the EU authorized representative is:
Springer Nature Customer Service Center GmbH
Europaplatz 3, 69115 Heidelberg, Germany

Printed by Libri Plureos GmbH
in Hamburg, Germany